This unique book ranges across the physical, biological and social sciences in the development of its primary theme, that there are nine major 'integrative levels' which can be recognised. The term integrative levels was first used by Joseph Needham in 1937, and has two key features. The first is that members of a given integrative level are unified entities and the second is that a member of one level is commonly composed of parts which are members of the next lower level. Thus fundamental particles form Level 1 while Level 9 is that of sovereign states. This theme has been developed by Max Pettersson in a book which explores the many links between the physical, biological and social sciences, reaching wide-ranging and sometimes unexpected conclusions.

T0297130

COMPLEXITY AND EVOLUTION

COMPLEXITY
AND
EVOLUTION

Max Pettersson

CAMBRIDGE
UNIVERSITY PRESS

CAMBRIDGE UNIVERSITY PRESS
Cambridge, New York, Melbourne, Madrid, Cape Town, Singapore, São Paulo, Delhi

Cambridge University Press
The Edinburgh Building, Cambridge CB2 8RU, UK

Published in the United States of America by Cambridge University Press, New York

www.cambridge.org
Information on this title: www.cambridge.org/9780521117951

First published 1996
This digitally printed version 2009

A catalogue record for this publication is available from the British Library

Library of Congress Cataloguing in Publication data
Pettersson, Max.
Complexity and evolution / Max Petterson
p. cm.
Includes bibliographical references and index.
ISBN 0 521 45400 X (hc)
1. Science.   I. title.
Q158.5.P47 1996
003'.7–dc20   96-15186 CIP

ISBN 978-0-521-45400-1 hardback
ISBN 978-0-521-11795-1 paperback

Contents

Foreword

The concept of integrative levels was one which I introduced, stimulated by various predecessors, in the Herbert Spencer Lecture which I gave at Oxford in 1937, over half a century ago.

Several dozen authors have subsequently used the idea of different integrative levels. Their discrimination of integrative levels has often used the 'compositional criterion'; but the present work by my old friend Max Pettersson is unique in also introducing a 'duality criterion', for the discrimination of major levels. One of the agreeable results of this is that we now find an excellent match between the relatively static hierarchy of things present today, and the evolutionary progression to more and more complex entities which has occurred on Earth during the past 4000 million years.

Having established a series of nine major integrative levels, various quantitative studies have now become possible for the first time. Perhaps most or all of the twenty or so newly discovered quantitative conclusions will be found to remain valid. For readers a generation or two hence, it will be very interesting to notice how much further such studies have developed and ramified.

Joseph Needham

Preface

This book, about the various sciences, is addressed primarily to students and general readers. Apart from my main duties of lecturing in biology at Brunel University in West London, for many years I gave short courses of lectures plus discussion on the fundamental findings of science. It was for groups of first-year undergraduates whose main disciplines varied widely. And the following pages lean heavily on that teaching experience. Hence almost any student or general reader is likely to find them quite intelligible.

The topics covered certainly range widely, from small fundamental particles up to the large and complex societies of sovereign states. There is also a brief outline of the main evolutionary developments since the beginning of the Earth, more than four thousand million years ago.

After having established, objectively, nine major levels of organisation, several new types of quantitative study become possible for the first time. These are reported in Chapters 10 to 15.

During the 1930s I studied at Cambridge, and then worked there on biological research. In 1940 I learnt that I was about to join the Royal Corps of Signals, in the Army. At that time I knew very little about telecommunications, though later, as it happens, I ran the War Office Wireless Demonstration Section, giving instructional talks and demonstrations to all ranks up to Brigadier.

However in 1940, to remedy my ignorance I went into Heffer's bookshop in Cambridge, and there, from the shelves, was able to purchase a couple of useful textbooks. One was the *Admiralty Handbook of Wireless Telegraphy*. This contains a striking down-the-page logarithmic diagram of wavelengths and frequencies, some 20 centimetres long. And it is that diagram which can be regarded as the mother and father of all the distinctive logarithmic diagrams which help to illustrate the present book.

Due to its wide coverage of the different branches of science, and to the relatively simple way in which it is written, it is felt that this book should make a substantial contribution to the public understanding of science.

Acknowledgements

I wish to express gratitude to some two dozen colleagues at Brunel University, for specialist information generously supplied. Those who stand out include Professor Stan Bevan, for chemistry, Professor Peter Macdonald, for mathematics, Dr Jim Nodes, for biochemistry, Professor Gordon Onions, for entomology, Dr Sue Smith, for microbiology and Dr John Warren, for physics. Several luminaries from other organisations have similarly assisted, including Dr J. Finch of the MRC Laboratory of Molecular Biology. Also the Cambridge professor of geology, Professor I. N. McCave.

Several publishers or authors have kindly given permission for the inclusion of line drawings which have already appeared elsewhere.

My son Mike has helped with regard to geology, prehistory and history, as well as with preparation of the typescript. My son Tony helped concerning rates of change. I am grateful to my wife Marie, and to my daughter Helen, for not complaining about the many hours spent working on this book, hours when I might have been doing something else.

M.P.

Chapter 1
Natural and other hierarchies

Some definitions

One of the definitions which a dictionary gives for the word 'entity' is 'a thing that exists'. A definition of 'hierarchy' is 'a body classified in successively subordinate grades'. This corresponds reasonably well with what follows. The word 'integer' means 'a whole'. An individual mammal is certainly a coherent entity, made up and functioning as a whole. It is an 'integrated natural entity', and a good representative of one of the major 'integrative levels' which are the primary theme of this book.

A car and a computer are not 'natural' but manufactured, or 'artificial'. In comparison with a heap of scrap metal, a car is certainly an integrated entity; but an integrated artificial entity. A pebble and a planet are each 'natural' entities. But compare a pebble with a live mouse, of about the same size. If the pebble is broken in two, the properties of the system do not change greatly. But if a live mouse is cut into halves the properties change catastrophically because the live mouse has such a highly complex and well-integrated structure, with properties to match. While a mouse can be thought of as an integrated natural entity, pebbles and planets (being rather low-grade aggregations of molecules and ions, etc.) could be regarded as 'aggregational' natural entities, a term that will not often need to be used.

Natural hierarchies

It is well known to students of chemistry and physics that molecules consist of atoms, and that atoms consist of fundamental particles. For instance a molecule of water consists of two hydrogen atoms, plus one oxygen atom; and each hydrogen atom consists of two fundamental particles, a proton and an electron. For molecules and atoms in general, this kind of hierarchy may

1

Complexity and evolution

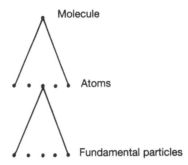

Fig. 1.1. Structural hierarchy leading up to a molecule. The dot at the apex of the figure represents one molecule and the five dots on the level below the two, several or many atoms of which the molecule is composed. One atom is then shown as composed of two, several or many fundamental particles.

be illustrated as in Fig. 1.1. Fig. 1.2 shows a longer hierarchy. This leads up to multicellular organisms: trees, butterflies, cats, dogs and humans, etc. The level below is that of 'ordinary' cells, with nuclei. The nucleus of a cell contains a number of chromosomes. Each is a string of genes. It is the chromosomes which send out biochemical messages which regulate the activities within the cell, and indirectly the activities of the whole organism.

Certain organisms, such as some viruses and bacteria, contain only a single chromosome. These are hence forward referred to as 'intermediate entities', since they are taken to be intermediate between the level of molecules and the level of ordinary cells with nuclei (the term intermediate entities was introduced in 1976). Since an ordinary cell contains a number of chromosomes, the cell can be regarded as being made up of a number of intermediate entities, each centred on a single chromosome.

From its beginning, the planet Earth contained molecules, atoms and fundamental particles. The first living things to appear were intermediate entities. Then came the larger and more complex cells with nuclei. Cells grow and then increase in number by cell division, one cell dividing into two cells. If the daughter cells did not go apart, after a number of divisions, and if some differentiation took place, there is the possibility of an entity at a higher level evolving – a multicellular organism. And this 'staying together after reproduction' has obviously occurred, independently, in quite a number of different lineages. (Further biological topics are discussed in Chapters 5 and 6.)

In Fig. 1.2, the various levels are numbered from 1 to 6. Further justification for this kind of numbering will be given in some detail in the next chapter. However, it appears that, on Earth, we now have entities that are members of nine different major levels. Levels 4 to 9 have evolved in

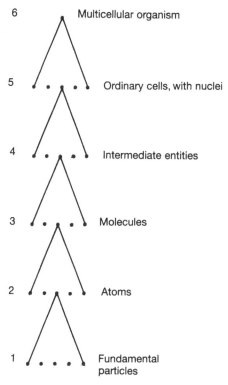

Fig. 1.2. Longer hierarchy leading up to a multicellular organism. These first emerged around a thousand million years ago.

sequence, as sketched in Fig. 1.3. In most species of animal the young (if they are mobile) go apart from the mother almost immediately they are produced; or the mother lays eggs and just leaves them. But about 200 million years ago some mothers began to produce milk, from special 'mammary glands', to feed their young – and became mammals. The mother and young stayed together, at least for a period. The 'one-mother family society' had arisen, an entity at level 7. This is a second instance of the emergence of a higher level entity through 'staying together after reproduction'. While this behaviour first occurred in mammals, it can also be seen in certain species of bird and insect.

Chapter 7 will discuss some cases where several one-mother family societies are integrated into a larger coherent group, a multifamily society at level 8. Examples include those of termites, whales and mountain gorillas.

Each entity of level 9 is the society that comprises a sovereign state. In Chapters 8 and 9 we shall pay special attention to human societies. As might be expected, one or two special problems arise. For instance it is found con-

Complexity and evolution

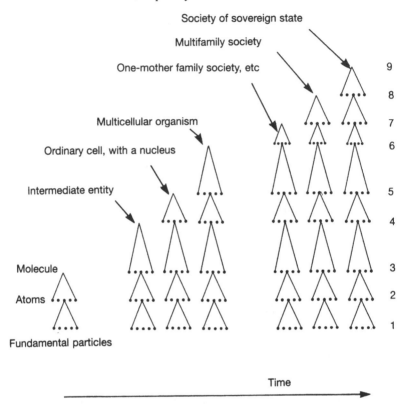

Fig. 1.3. A summarising diagram about evolution on Earth. The seven hierarchies have been arranged along a non-quantitative axis of time; and the levels have been numbered as in accordance with the criteria of Chapter 2.

venient to regard a 'one-kitchen household' as an alternative kind of member of level 7. The society of a human settlement is a multifamily society, at level 8; and a settlement may be defined on either geographical or political criteria. Then at the peak of the hierarchy, at level 9, we have the 'society of a sovereign state'. One refers to the persons concerned including their inter-relationships.

Other hierarchies

In many countries, both the Army and the Church exhibit hierarchical structures (Figs. 1.6, 1.7). Our cultural heritage has been much enriched by the Christian Church. One of its useful offerings is the word hierarchy itself. The term 'hierarch' means 'a ruler in holy things', or chief priest, being

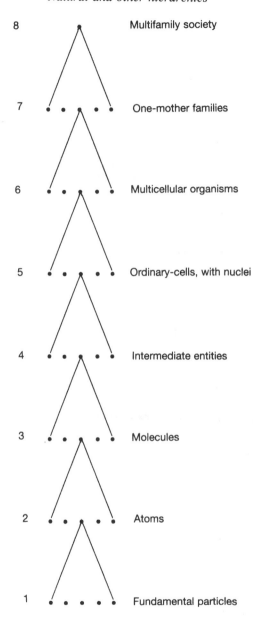

8   Multifamily society

7   One-mother families

6   Multicellular organisms

5   Ordinary-cells, with nuclei

4   Intermediate entities

3   Molecules

2   Atoms

1   Fundamental particles

Fig. 1.4. Around 200 million years ago, social evolution commenced. There were at first one-mother family societies, and then later also multifamily societies.

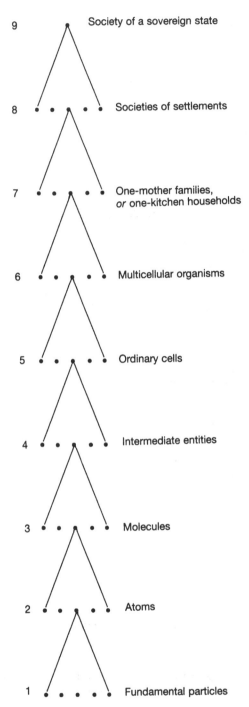

9   Society of a sovereign state

8   Societies of settlements

7   One-mother families,
     or one-kitchen households

6   Multicellular organisms

5   Ordinary cells

4   Intermediate entities

3   Molecules

2   Atoms

1   Fundamental particles

Fig. 1.5. Finally the societies of sovereign states emerged, around eight thousand years ago.

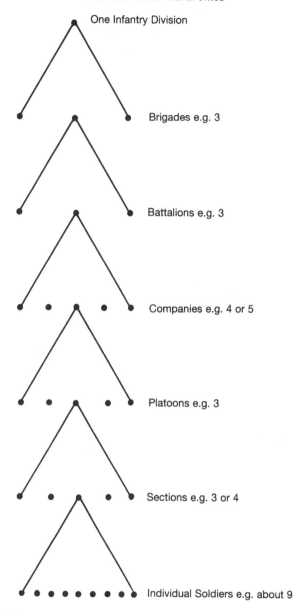

Fig. 1.6. The hierarchical structure of a British Infantry Division. Certain possible extras and variants are omitted. (Compiled from various sources, including a British government publication *The British Army* (1987).)

Complexity and evolution

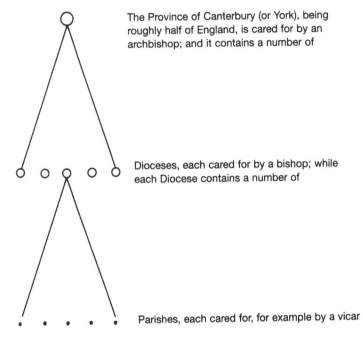

The Province of Canterbury (or York), being roughly half of England, is cared for by an archbishop; and it contains a number of

Dioceses, each cared for by a bishop; while each Diocese contains a number of

Parishes, each cared for, for example by a vicar

Fig. 1.7. A simple version of the hierarchical organisation of the Church of England. Besides referring to the areas of land concerned, there is a corresponding hierarchy of church officials.

derived from the Greek words *heiros*, sacred, and *archein*, to rule. Indeed the word hierarchy was originally used for the entire body of clergy, classified in successively subordinate grades (compare Fig. 1.7). The Pope could be described as the hierarch of Catholic Christendom.

Chapter 2
Major integrative levels

Ideas about integrative levels

It was Joseph Needham who introduced the term integrative levels. This was at Oxford in 1937, during a lecture in memory of the philosopher Herbert Spencer (1820–1903). For the purpose of brief statement, he found it convenient to recognise the series of levels of organisation shown in Table 2.1.

The use of the adjective 'integrative' can be taken to imply two different points: (a) that the members of a given integrative level or category are themselves unified or integrated entities, or wholes; and (b) that a member of one level is commonly composed of, or integrated from, parts which are themselves members of the next lower level. It can be seen that in general the series in Table 2.1 is consistent with both these points.

A number of authors have taken up the idea of a series of integrative levels, besides the present writer; and their publications have been reviewed by Jolley (1973), Foskett (1978) and Pettersson (1978b) among others.

Table 2.1. *Scale of levels of organisation, or integrative levels, cited by Needham*

Social association
Animal body
Organ
Living cell
Colloidal aggregate
Molecule
Atom
Ultimate particle

Joseph Needham

Needham was born in 1900 and had a deeply religious upbringing. During the 1920s and 1930s he helped to establish and develop the new subject of biochemistry at the University of Cambridge and completed the standard work *Chemical Embryology*. During the Second World War he spent much time in war-torn China, on behalf of the British Government; and while there learnt Chinese. He helped to launch the United Nations Educational, Scientific and Cultural Organization (Unesco), as first head of the S or science department. He has been a Fellow of the Royal Society since 1941. A book by Wersky (1978) provides a collective biography of a group of distinguished scientists, including Needham, who were notably concerned with social as well as scientific problems in the interwar years.

After the Second World War Needham established at Cambridge the East Asian History of Science Library, as well as the Needham Research Institute. And he was actively engaged, in collaboration, on the immense work *Science and Civilisation in China*. This already runs to nearly 10 000 pages. He also served for a period as Master of his Cambridge college, Gonville and Caius. He died in 1995.

The more important integrative levels

It is more than 50 years ago that Needham suggested the eight levels listed in Table 2.1. What I myself now regard as the nine *major* levels are set out in Table 2.2. For instance, the level of 'social associations' has been divided into three. This will be discussed in Chapter 7. Level 6, that of multicellular organisms, is not now restricted to animals, but includes also green plants and fungi such as the mushroom. Level 4 is that of intermediate entities, each centred on a single chromosome. And the level of biological organs has now been omitted, as not really being a major level.

It is interesting to compare and contrast the level of ordinary cells and that of biological organs. As Fig. 2.1 indicates, some 'ordinary' cells collaborate as part of a multicellular organism, while others such as *Chlamydomonas* and amoeba, both of which flourish in fresh water, exist quite independently as single-celled living organisms. *Chlamydomonas* is a genus of green algae some of which swim by using two flagella, while amoeba is an animal which moves about on solid surfaces. On the other hand, any biological organ such as a heart or a leaf does not flourish independently, under natural conditions, while still alive. Organs only collaborate, within multicellular organisms, while cells have a dual role: some exist independently while others do

Table 2.2. *The integrated natural entities which are members of the nine levels, and the three ranges in the natural hierarchy*

Major integrative level	Members of levels	Range of discipline
9	Societies of sovereign states	
8	Multifamily societies	Social range
7	One-mother family societies etc.	
6	Multicellular organisms	
5	Ordinary cells, with nuclei	Biological range
4	Intermediate entities, each centred upon one chromosome	
3	Molecules	
2	Atoms	Physical range
1	Fundamental particles including (or together with) photons	

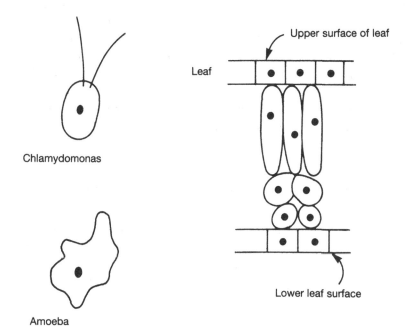

Fig. 2.1. Some cells, independent or otherwise, much magnified but not drawn to scale. In each cell there is a nucleus, shown shaded, containing a number of chromosomes.

collaborate. This can be taken to indicate that the level of cell is a major level, while that of biological organ is not.

Two special rules or criteria

It has become clear to me that there are two main criteria by which the nine major integrative levels have come to be discerned and discriminated:

(1) By the *compositional criterion*, it is a requirement that each entity of any major integrative level (except the lowest) materially consists mainly of entities of the next lower level.

(2) By the *duality criterion*, it is a requirement that *some* of the entities of any major integrative level (except the highest) are joined, bonded or fused together with others, with which they collaborate in the constitution of an entity of the next higher level, while *some other* entities of the same major level exist as free and independent entities.

The situation is well illustrated by the middle level in Fig. 2.2.

By the compositional criterion, a whole normally consists 'mainly' of parts which are members of the next level below. But a word such as 'mainly' is necessary: for instance, in a horse or a person, the blood plasma exists between cells rather than within cells. A multicellular organism consists *mainly* of cells, but 4 per cent of a person's body weight is made up of the extracellular molecules, etc. of blood plasma (Fig. 2.3); and there are other similar instances.

Therefore, the series of nine major integrative levels which I use here, are at least dignified by having had two precise rules or criteria applied during their discrimination.

The evolutionary ascent from Level 3 to Level 9

Several authors besides Needham have listed biological organs as one of the main integrative levels. However, if organs are excluded, as consistent with the duality criterion, it is now possible to construct Fig. 2.4. This shows the approximate dates at which entities of Levels 4, 5, 6, 7, 8 and 9 emerged, in succession, through the innovative collaboration of entities at the next level below.

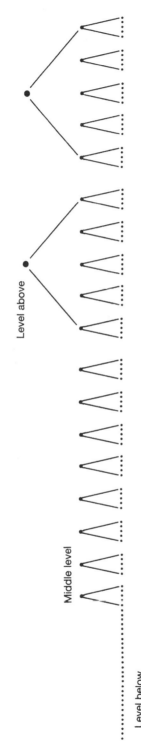

Fig. 2.2. The middle level of entities does conform both to the compositional criterion and to the duality criterion.

Complexity and evolution

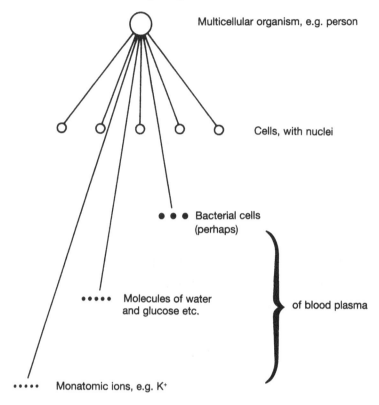

Fig. 2.3. This shows some of the extra components of which a multicellular organism may consist, components which are at a lower level than the cells mainly constituting the organism, and which are between rather than within the cells.

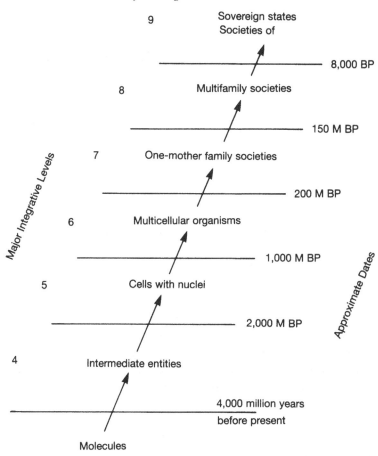

Fig. 2.4. A diagram illustrating the evolutionary progression from entities of one major integrative level to entities of the next level above. This diagram would not have been possible if for instance the level of biological organs had been regarded as a 'major' level. (M BP, million years before present.)

Chapter 3
Some logarithmic forms of display

By now we have covered all the more general points involved in the idea of the natural hierarchy, and of its nine major integrative levels. A number of points of detail will need to be filled in, as in Chapters 4–9, before the quantitative results of Chapters 10–14 can be attained. But before discussing atoms and molecules in Chapter 4, it seems appropriate – for the benefit of non-specialist readers – to give some idea of how large such things as atoms and molecules actually are, and how their sizes compare with those of viruses and cells. At the same time we can briefly introduce some methods of logarithmic plotting which are rather useful for the display of the weight (or mass) of various entities, and their parts, as well.

Lengths of smaller entities

Starting from the left-hand end of a ruler, each movement to the right by a standard distance (say an inch or a centimetre) represents an additive increase of length by that standard unit. But the vertical scale in Fig. 3.1 is arranged differently. Each movement upwards, from one peg to the next peg above, represents a multiplication by a constant factor, in fact by 10, thus the distance on the scale between 1 millimetre and 1 centimetre is the same as that between 1 nanometre and 10 nanometres. We shall be using this kind of logarithmic scale frequently, so for those who wish for it, a little more information about the metric scale of lengths, and logarithmic plotting, is offered in an Appendix to this chapter.

It is well known that there is an enormous range in length between the smallest and largest multicellular organisms: there are tall forest trees, and great whales, while some others are too small to be seen (easily) with the naked eye. There is an enormous range in length between the shortest and longest 'ordinary' cell (with a nucleus) too. After all, the thin extension of

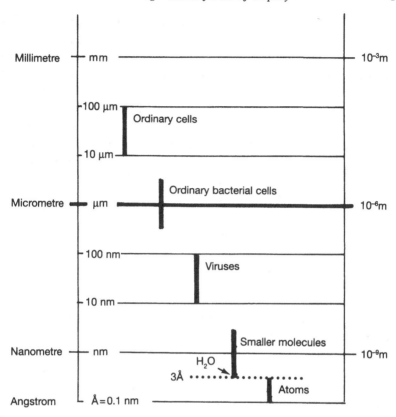

Fig. 3.1. A *rough* guide to the *main* ranges of lengths of various smaller entities. Some may prefer to regard this as a rough guide to linear dimensions in general, i.e. to breadth, etc. as well as length.

a single nerve cell goes the whole length of a limb, and a giraffe, for instance, has very long limbs. But a considerable proportion of all ordinary cells *are* between 10 and 100 micrometres in length, as Fig. 3.1 reminds us. The other kinds of entity mentioned there, however, have very much smaller ranges of length. (See Pettersson, 1964, for a similar diagram, but containing much more of the total information.) Fig. 3.1 on the other hand is intended only as a rough guide to typical lengths. Some of the main ranges, as drawn, terminate at the level of one or other of the pegs along the vertical axis (that is at 10^{-n} metre, where n is a whole number). For example, the range for ordinary cells is given as from 10 to 100 micrometres. Many cells of ordinary bacteria are one to several micrometres long, and some are as small as a third of a micrometre. So in this diagram bacteria are characterised as typically from about a third of a micrometre in length, to 3 micrometres. (The half-way point between the levels of 1 micrometre and 10 micrometres corresponds in fact to the square root of 10, or 3.1623 micrometres; but both 3 and 3.33 are useful approximations.) The diagram shows viruses as typically between a tenth and a hundredth of a micrometre in length. So for the first three entries – those for ordinary cells, ordinary bacterial cells and viruses – there is a useful symmetry around the level of micrometre.

Turning now to consider molecules, each of us in our chromosomes contains molecules of DNA which would apparently be more than a millimetre long if straightened out. And a diamond can be ranked as a molecule (see Chapter 4). But these are quite exceptional molecules. Certainly 'smaller molecules' may be said to range from about 3 angstroms to 3 nanometres in length. The typical size of a molecule of water – one of the smallest molecules – is about 3 angstroms. Thus, as Fig. 3.1 shows, the length of a small virus is only about 30 times that of a very small molecule, such as the water molecule. Compared with the other entities being discussed, there are only a relatively small number of different kinds of atom; and they happen to have a very small range of size. Many advanced textbooks cite the radii of different atoms, in a variety of different situations, and sometimes one can think of an atom as a sphere, and then its length is equal to its diameter. It is found that very many atoms, from hydrogen to uranium, are in fact between 1 and 3 angstroms in length (or diameter). For instance an aluminium ion Al^{+++}, which has lost three electrons, is 1.0 angstrom across; a carbon atom (with covalent single bonding all round) is 1.544 angstroms across; while a uranium atom, in metallic uranium, is 2.85 angstroms.

Mass versus size

We have been discussing the lengths of entities. Unfortunately when one gets down to the level of electrons and photons, there is no satisfactory information available about such properties as their length, breadth and volume. On the other hand one can specify the weight or mass of an electron, and the mass of each different kind of moving photon. *Hence when reviewing entities throughout the hierarchy, one has to rely on the property of mass, rather than on the property of size.*

Special triangles

For display purposes it is sometimes interesting to compare the mass of an integrated natural entity, with the range of mass of its various parts, which are members of the integrative level below. In the next chapter we will do

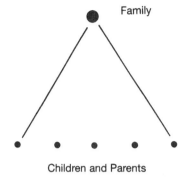

Fig. 3.2. Hierarchical diagram of a human family, consisting of parents and children.

this for the electrons, protons and neutrons that make up the atom, but here the method can be illustrated by the human family. As an example, consider a family of two parents with three children. The whole, consisting of five parts, can be represented as in Fig. 3.2. In considering their weight or mass, it will be convenient to use the model of a family as shown in Fig. 3.3, where the father weighs 100 kilograms, the baby 10 kg, and where the total for the five individuals adds up to 300 kg. The two lower corners of the triangle show the weights of the lightest and heaviest members of the family, while the apex shows the family's total weight. Such triangles will be used frequently; indeed it is convenient to think of them as special gravimetric triangles. In this diagram, by the way, only the vertical axis is quantitative, with the items plotted higher or lower according to their mass, the horizontal arrangement is, however, merely a question of artistic convenience.

The material composition of each member of the family includes various organs and tissues. There in turn are composed of cells, each of which arose from a previously existing cell by cell division. Each cell can be regarded as composed of intermediate entities, which are composed of molecules, composed of atoms, composed of fundamental particles. It would be agreeable to display the weights of these entities by using a set of triangular diagrams which showed the family composed of individuals, then one individual as composed of cells, and one cell as composed of intermediate entities, and so on. This is obviously impossible in Fig. 3.3, since there is such a small distance on the vertical axis, between 0 and 10 kilograms, indeed, even the 1 gram level would be scarcely above the zero baseline.

Fortunately the device of logarithmic plotting can come to our aid. Fig. 3.4 shows our family model arranged in this different way. On such a scale, of course, there is as much distance between the 1 and 10 gram levels as

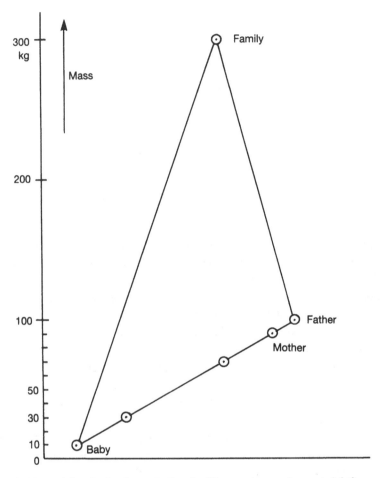

Fig. 3.3. The weight of a family, and of each of its members, using a straightforward arithmetic axis of mass.

there is between 1 and 10 kilograms. If the axis were extended there would similarly be as much space between 1 and 10 micrograms, and 1 and 10 nanograms, and so on. In the next chapter we shall be able to plot – along a logarithmic axis of mass – one molecule as composed of atoms, ranging from lightest to heaviest atoms; and then one atom as composed of its lightest to heaviest fundamental particles. An entire natural hierarchy can thus be plotted quantitatively, using a series of special gravimetric triangles, along the same kind of axis (see Fig. 9.2).

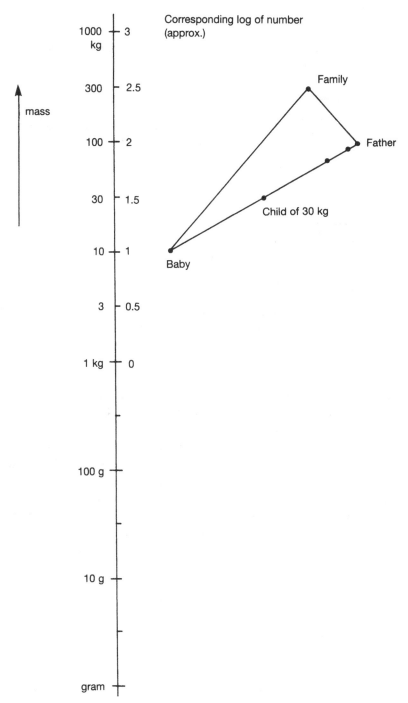

Fig. 3.4. Another plot of the same family model, now using a logarithmic axis of mass.

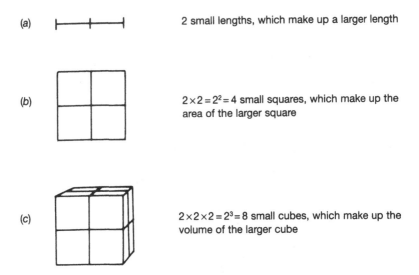

(a)   2 small lengths, which make up a larger length

(b) $2 \times 2 = 2^2 = 4$ small squares, which make up the area of the larger square

(c) $2 \times 2 \times 2 = 2^3 = 8$ small cubes, which make up the volume of the larger cube

Fig. 3.5. An illustration of the relationships between linear dimensions, area and volume. If the side of a square is doubled, its area increases fourfold; while if the side of a cube is doubled, its volume increases eightfold.

Length, volume and mass

There is a further form of logarithmic display which is worth mentioning before dealing with the very small entities discussed in the following chapter. Many of us, at some time or other, will have encountered the relationships illustrated in Fig. 3.5. If one considers a number of small units of length (in this case two; Fig. 3.5a) which together make up a greater length, then the corresponding number of small areas which make up the larger area is greater (Fig. 3.5b). It is 2×2 or 4. And the corresponding number of small cubes, which make up a larger cube, is much greater. In the present example (Fig. 3.5c) there are 8 small cubes in the larger cube. The number is $2 \times 2 \times 2$ or 2^3. Or one can say that it is the original number raised to the power 3.

Earlier in the chapter we saw that the micrometre was a unit of length which was convenient for describing the linear dimensions of bacteria. There are 10^4 micrometres in a centimetre, but there are $10^{4 \times 3}$ or 10^{12} cubic micrometres in a cubic centimetre. Another way is to say that there are a million million cubic micrometres in a cubic centimetre. Similarly there are 10^8 angstroms in a centimetre, but $10^{8 \times 3}$ or 10^{24} cubic angstroms in a cubic centimetre.

The founding fathers of the metric system, in the 1790s in France, decided to use the ubiquitous substance water for helping to obtain a unit of mass. Having already decided on units of length, they defined the gramme (or gram) as the weight of one cubic centimetre of water. Hence it is convenient to have the scales aligned so that cubic centimetres and gram are on the same level. And one can now describe various weights as the weights of cubes of water which – at least in some cases – can be fairly easily pictured in one's mind.

Some bacterial cells are about a cubic micrometre in volume, this being a million-millionth of a cubic centimetre. Their density is little different from that of water. So corresponding, the mass of such a bacterial cell is about one picogram, or a million-millionth of a gram.

If we are to get to the stage of plotting the whole natural hierarchy quantitatively – as a series of special gravimetric triangles along a logarithmic scale of mass – it will be necessary to be able to cite the mass of atoms and molecules also in metric terms. This is done, so far as is necessary, in Chapter 4. (See Pettersson, 1964, for further information on the metric mass of atoms, etc.)

A hydrogen atom weighs a little more than a million-million-million-millionth of a gram; a little more than 10^{-24} gram. But to refer to that unit as 'ten to the minus twenty-four gram' is certainly rather cumbersome. However, we can now use the new term 'yoctogram' (see Appendix). The level of 10^{-24} gram is still rarely referred to, as specialists use the traditional atomic mass unit, which is now defined by taking the mass of an atom of carbon-12 as being 12 atomic mass units (or 12 a.m.u. or 12 daltons). It so happens that one atomic mass unit is just about 1.66 yoctograms. Thus the metric mass of a carbon-12 atom is about 20 yoctograms, that of a water molecule is about 30 yoctograms, while a glucose molecule is roughly 300 yoctograms.

Postscript

It should be pointed out that no disrespect is intended when one considers the weight of an individual person or of a social group. Though for a certain purpose one may plot their weights, one can remain fully aware of their other important, complex and fascinating characteristics. In the same way chemists imply no disrespect to the fascinating properties of mercury, gold, carbon or oxygen, when they consider the weights of the atoms of these elements. It was only when John Dalton and others paid particular attention to the weights and volumes of various chemical substances that the existence of atoms and molecules became properly established. Similarly, by considering the weights of integrated natural entities, we can obtain a much more precise and quantitative understanding of the whole natural hierarchy than would otherwise be possible. Fig. 1.5 gives an introductory view of its structural organisation, but Fig. 9.2 for instance will contain much more information.

Appendix to Chapter 3

There are two topics to be dealth with:

The metric system
Logarithms and logarithmic plotting.

The metric system

One ancient unit of length was the cubit, being the distance from a person's elbow to the tip of the middle finger, but that varied from 18 to 22 inches. During the eighteenth century there were several dozen different systems of weights and measures in Europe alone. The long-pound of Sweden was more than twice as heavy as the short-pound of Venice. It was all very confusing for trade and commerce. There were even three different pounds in London, the much-used avoirdupois pound and two different troy pounds.

After the French Revolution of 1789, the new French government decided to do something about it, and by 1800 had set up the metric system. This has eventually become internationally accepted, at any rate for scientific purposes.

The French decided to base a new unit of length upon some feature of the planet Earth, upon which all peoples live. They obtained a good estimate of the distance (at sea level) from the North Pole to the Equator. They then divided this by 10, again and again, until they got something like our yard and called that the length of a *metre*. The word comes from the French and Greek words for 'measure'. The French used the distance between two marks on a special metal rod (at a certain temperature) as the standard example of a metre; and copies were distributed. Following earlier suggestions, derived units were all separated by one or more factors of 10. Thus a hundredth of a metre is a centimetre.

To obtain a new unit of weight (or mass) the ubiquitous substance water was used. The weight of a cubic centimetre of water (just above freezing point) gave the new unit of weight, the gramme or gram. The name comes from the Latin word *gramma*. In Roman times this stood for a small unit of weight, which was in fact not very different from the gram.

For many years, the centimetre, the gram and the second were regarded as the central units: the CGS system, but since 1960 attention has focussed on the metre, the kilogram and the second: the MKS system. MKS units are now often referred to as SI units, using the initial letters of the French term Système International.

One can read much more about the metric system in, for instance, a book by Danloux-Dumesnils (1969). Among other things it mentions is that the

Table 3.1. *Metric prefixes and symbols*

Factor	Prefix	Symbol
10^{24}	yotta	Y
10^{21}	zetta	Z
10^{18}	exa	E
10^{15}	peta	P
10^{12}	tera	T
10^{9}	giga	G
10^{6}	mega	M
10^{3}	kilo	k
10^{2}	hecto	h
10^{1}	deca	da
10^{0}	-	-
10^{-1}	deci	d
10^{-2}	centi	c
10^{-3}	milli	m
10^{-6}	micro	μ
10^{-9}	nano	n
10^{-12}	pico	p
10^{-15}	femto	f
10^{-18}	atto	a
10^{-21}	zepto	z
10^{-24}	yocto	y

metre is now redefined as so-many times the wavelength of a certain natural radiation.

Table 3.1 gives an official list of all the metric prefixes and symbols, as published in 1991 by the National Physical Laboratory.

Logarithms and logarithmic plotting

A hundred is 10×10 or 10^2, so the logarithm or 'log' of a hundred is 2. Similarly the log of a thousand, or 10^3, is 3. And the log of a million is 6.

Logarithmic scales are used in diagrams throughout this book. A few examples follow of the numbers (to four significant figures) which correspond to certain logarithms.

Log	*Number*
2	100
1.8	63.10
1.5	31.62
1.3	19.95
1	10

Chapter 4
Physical range of integrated natural entities

Introduction

As indicated in Chapter 1, the first three major integrative levels are those of fundamental particles, atoms and molecules. These constitute the physical range of integrated natural entities.

A hydrogen atom consists of two fundamental particles. Most of any atom's mass resides in its small central nucleus. The nucleus of a hydrogen atom consists of a single proton. This has a positive electrical charge, which counterbalances the negative charge of the electron which circles or orbits around the nucleus (Fig. 4.1a). Each atom of the other chemical elements has more than one electron orbiting the centre. And apart from hydrogen atoms, the central nucleus of all normal atoms contains two or more protons together with two or more neutrons. Neutrons have a mass similar to that of protons, but they are electrically neutral, hence their name.

When a sunbeam enters a room one can speak of the arrival of electromagnetic waves, of such and such wavelength and frequency. Or alternatively one can speak of tiny particles streaming in, which have been given the name photons. In some contexts photons are not classed as fundamental particles, but in our new classification photons are definitely regarded as members of Level 1; so it is convenient here also to regard them as fundamental particles. Physicists sometimes regard photons as having no weight or mass; or no mass when not moving. But when moving, as in a beam of light, each photon does have a mass which can be calculated from knowing the wavelength or frequency of the light concerned. And since we shall often wish to compare the masses of various entities, it is in their moving condition that we shall normally refer to photons. When a photon arrives at a light-absorbing surface its energy (plus its attendant mass) goes into the surface, and the photon itself is no more. Besides visible light, the other kinds of electromagnetic radiation, such as X-rays, ultraviolet, infrared, and radiowaves, can similarly be thought of as streams of photons.

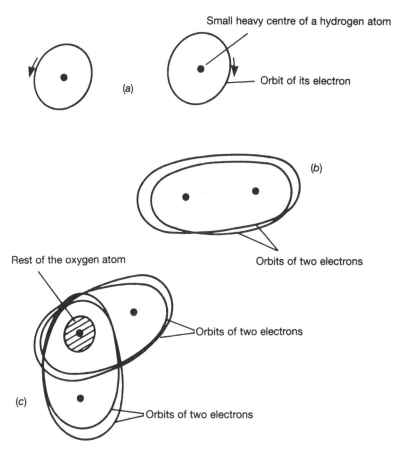

Fig. 4.1. Schematic diagrams of (*a*) two free hydrogen atoms, (*b*) a hydrogen molecule, and (*c*) a water molecule, H_2O (see text).

A number of other kinds of fundamental (or elementary) particles have been discovered, in recent years, and perhaps it would be useful briefly to review the range. Apart from photons there are two main families of fundamental particles: (1) baryons, the heavier ones, including the proton and the neutron; and (2) leptons, the lighter ones, including the electron, neutrino and muon. There are also some rather transitory particles usually of medium mass, the mesons. In addition there sometimes occur particles such as the antiproton which has the same mass as the proton but has an opposite, negative, charge.

It should be further mentioned that a proton, for instance, can be considered as consisting of three quarks. Quarks do not themselves rank as fundamental particles, and so far they have not been detected outside the particles of which they form a part. Of course future research may suggest replacing the present major integrative Level 1, by two Levels, say Level 0 and Level 1. At the other end of the scale future social evolution may, similarly, see the emergence of an entity at Level 10.

Mass of fundamental particles, atoms and molecules

Each proton and neutron weighs about 1.67 yoctograms. But an electron weighs only about a thousandth of a yoctogram, or 0.000911 yoctograms. Or about 10^{-27} grams. The mass of most photons, however, is much smaller still and depends upon the wavelength of the electromagnetic radiation concerned. Thus, if an electromagnetic wavelength is expressed in metres, then that number, divided into 2.2×10^{-39}, gives the photon's mass in grams. That is:

$$\text{mass of photon (in grams)} = \frac{2.2 \times 10^{-39}}{\text{wavelength (in metres)}}$$

The above data on the mass of photons will be useful when we come to review the range of mass of all the entities of each of the nine major integrative levels in Chapter 13.

Table 4.1 gives some approximate mass values of various fundamental particles, atoms, etc. both in metric terms and in the normally used atomic mass units.

During 1989 scientists at CERN, the European Laboratory for Particle Physics, announced the mass of the very transient Z° particle. At 162.3×10^{-24} gram it is a quite exceptionally heavy fundamental particle, having nearly a hundred times the mass of proton.

Many years ago chemists introduced a splendid system for referring to the *relative* weights of atoms of the different elements, long before the *absolute* weights were known. The weight of the lightest atom, that of hydrogen, was taken as unity, i.e. as 1 atomic mass unit (a.m.u.), or (later) as 1 dalton (after John Dalton who developed the modern atomic theory of chemistry in 1808).

Taking the weight of hydrogen as 1 a.m.u., then the weight of a carbon atom is almost exactly 12 a.m.u., and that of an oxygen atom 16. For various practical reasons, however, it is the ordinary carbon atom, the ^{12}C atom, which is now taken to weigh exactly 12 a.m.u. This gives the hydrogen atom a weight slightly greater than unity (Table 4.1). The atomic mass unit now equals about 1.66 yoctograms, or, to six figures, 1.660 54 yoctograms.

A further point may be mentioned. The number of ^{12}C atoms in 12 grams of carbon-12 is known as the Avogadro number, or Avogadro constant. Over the years it has become estimated more and more accurately. It is reported

Table 4.1. *Some approximate mass values of certain physical entities.*
(One yoctogram = 10^{-24} gram.)

Molecule, atom or fundamental particle, etc.	Approximate mass	
	In metric units	In atomic mass units
Haemoglobin molecule	110 000 yoctograms	66 000 amu
Uranium atom	395 yg	238
Glucose molecule	299 yg (cp. 300)	180
Sulphur atom	53 yg	32
Sodium atom	38 yg	23
Water molecule	30 yg	18
Oxygen atom	26.6 yg	16
Nitrogen atom	23 yg	14
Carbon-12 atom	20 yg	12.000
Helium atom	6.6 yg	4
Neutron	1.674 9 yg	1.008 66
Ordinary Hydrogen atom	1.673 5 yg	1.007 83
Proton	1.672 6 yg	1.007 28
Atomic mass unit	1.660 540 yg (or 1.66 yg)	1
Electron	0.000 911 yg	0.000 549
Photon in motion in:		
blue-green light with 0.5 μm wavelength,	4.4×10^{-33} gram	
red light with 0.7 μm wavelength,	3.2×10^{-33} g	
radio waves of 330 m wavelength; or	6.7×10^{-42} g	
associated with		
alternating current of 50 hertz.	3.7×10^{-46} g	

that the value recommended recently by the Royal Society, in conjunction with the Institute of Physics and the Royal Society of Chemistry, is 6.022 136 7±0.000 003 6×10^{23} Thus the Avogadro constant is a little less than 10^{24}, which corresponds with the atomic mass unit being a little more than 1 yoctogram. Indeed one derives the value (mentioned above) of the atomic mass unit in yoctograms, by dividing the six-point-something part of the Avogadro constant into ten.

It was in 1811 that Avogadro, in what is now Italy, suggested that equal volumes of different gases, at the same temperature and pressure, contained equal numbers of gas molecules. At first ranked only as a hypothesis, by about 1860 it could be accepted as Avogadro's law (though the word hypothesis did continue to be used for many years). The Avogadro number of gas molecules, at 0 °C and at a pressure of one atmosphere, occupies 22.4 litres. That is an often-quoted volume. But multiplying 22.4 by 1.66 gives 37.2

Complexity and evolution

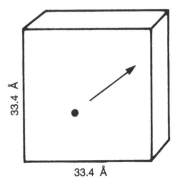

33.4 Å

Fig. 4.2. With 1 millimetre representing 1 angstrom, this cube represents the 'occupation volume' of one molecule of gas (at 0 °C and one atmosphere). It might be a molecule of hydrogen, an angstrom or so long. The arrow is to remind one that the molecule is in rapid motion.

litres, which will be the volume occupied by 10^{24} gas molecules. And this information makes it much easier to think down to molecular magnitudes. A litre would be three pegs above the level of cubic centimetres. So if 10^{24} gas molecules occupy 37.2 litres, then the average 'occupation volume' of a single molecule will be 37.2 cubic nanometres. Or the volume of a cube of which the side is 3.34 nanometres, or 33.4 angstroms (Fig. 4.2).

From Table 4.1 we may notice, among other things, the relatively small range of mass between the lightest atom, that of hydrogen, and the heaviest atom of any naturally occurring element, that of uranium. They all lie between 1 and 400 yoctograms.

Isotopes and isotopic dating

The nuclei of atoms are composed of protons and neutrons. The number of protons is characteristic of the particular element and is identical with its atomic number. It is equal to the number of extra-nuclear electrons, which determines the element's chemical behaviour. Although the number of protons in the nucleus of any particular element is fixed, the number of neutrons is not. Many elements have atoms with different numbers of neutrons in the nucleus; these atoms have different masses and are known as isotopes.

The normal carbon atom has six electrons and, in its nucleus, six protons and six neutrons. From the total content of the nucleus it is referred to as a carbon-12 (or ^{12}C) atom.

While most of the carbon dioxide in the air contains ^{12}C atoms, there is a small proportion with ^{14}C atoms. These decay at a steady rate, unaffected by temperature or pressure. Half the ^{14}C atoms decay in 5570 years. One says that the 'half-life' of ^{14}C is 5570 years.

The organic compounds in the trunk of a tree are made partly from carbon dioxide, and thus will at first contain the same proportion of ^{14}C atoms as the air. But as they decay their proportion decreases, and this enables archaeologists to assess how long ago the trunk was formed and one would expect this to be a useful method of 'isotopic dating'. However, after counting the annual rings in a series of tree trunks, it was found that the date estimated by ^{14}C is sometimes several centuries out (Renfrew 1971). The reason is presumably that the proportion of ^{14}C in the air has not remained constant.

For the dating of rocks which are millions of years old, geologists have used uranium, besides a number of other elements including thorium, rubidium and potassium among others. New methods and refinements are being continually introduced. There are a number of isotopes of uranium. The commonest (present as over 99%) is ^{238}U, which has a half-life of about 4500 million years; another isotope, ^{235}U, has a shorter half-life of about 700 million years. By assessing the amount of each isotope remaining in a portion of rock, and the amount of the breakdown products, one can tell how long ago the material welled up and solidified at or near the surface of the Earth's crust.

It is reassuring that the dating methods which depend on different chemical elements normally do confirm each other.

Major integrative Levels 1, 2 and 3

Let us check that the members of these three Levels do satisfy the compositional and duality criteria discussed in Chapter 2. Being themselves at the bottom Level (so far as is known) one cannot expect all fundamental particles to be composed of members of a still lower Level. But they do have an appropriate duality of behaviour; for some collaborate to make up atoms (of Level 2), while some electrons and all photons are free and independent.

As for the atoms of Level 2, during this century it has become known that they are indeed made up of members of Level 1. And they clearly show duality of behaviour. Atoms of the elements of helium, neon, argon, krypton, xenon and radon commonly remain free and independent; while atoms of carbon and oxygen and many other elements commonly collaborate to form entities of a higher Level.

Then consider the molecules at Level 3. The molecules of hydrogen, water, glucose, haemoglobin and the silicates all consist of two or more atoms. However there is also duality of behaviour. Some molecules of water and glucose for instance collaborate in the constitution of small living entities at

Table 4.2. *Short description of the members of the three major integrative*
levels of the physical range.

Major integrative level	Integrated natural entities which are members of the level concerned
Level 3	*Molecules*, being distinctive groups of two or more atoms, often held together by covalent bonds.
Level 2	*Atoms* (free or bonded, uncharged or charged) each being an integrated group of two or more fundamental particles.
Level 1	*Fundamental particles* include photons in motion as well as electrons, protons, neutrons.

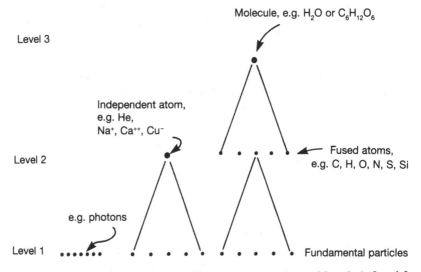

Fig. 4.3. Short structural hierarchies illustrating the members of Levels 1, 2 and 3.

Level 4, while at the same time very many water molecules and all the various silicate molecules in rocks do *not* take part in such collaboration.

Table 4.2 gives short descriptions of members of each of the three levels of the physical range. Fig. 4.3 shows some hierarchies, and helps to demonstrate conformity to the duality criterion.

Major integrative Level 3

Fig. 4.1*a* shows two separate hydrogen atoms. The diagram (*b*) shows a hydrogen molecule, H_2, where the two electrons orbit around *both* the atomic nuclei. With this arrangement one says that the two atoms are held together by a 'covalent bond'. Fig. 4.1c shows a water molecule, H_2O. Each hydrogen atom is joined to the oxygen atom, similarly, by a covalent bond. (Two electrons orbit around one hydrogen nucleus and the main part of the oxygen atom; and the same happens for the other hydrogen nucleus.)

In the case of both hydrogen and water, we have molecules held together by covalent bonds. These are molecules in the strictest sense of the term. However, the convention still persists of regarding common salt, NaCl, as consisting of molecules, despite the fact that the electrically charged pair of atoms (Na^+ and Cl^-) hardly ever exist as a discrete pair. In a similar way, many of the acids, bases and salts of school-time chemistry, also consist only of 'conventional' molecules. However a *weak* acid, such as acetic acid, does mainly exist as complete molecules.

Details of the various ways in which atoms cohere have only been discovered since 1900. Linus Pauling's standard work, *The Nature of the Chemical Bond*, was published in 1939 and was dedicated to Gilbert Lewis who was the first to draw attention, in 1916, to what is now called the covalent bond. Near the beginning of his book, Pauling notes that there are three main types of bonds between atoms, besides some intermediate types. And these three types may be briefly described. They are the metallic bonds, electrostatic bonds, and covalent bonds.

Metallic bonds

Such bonds hold together any piece of metal: a coin, a copper wire, or a metallic hull plate of an ocean liner. The outer electron of each metal atom becomes detached (or sometimes two do), so that the metal atoms are left as positively charged ions, among which the negatively charged electrons roam about freely. The opposite charges of the two kinds of entity hold the metal object together. And since an electric current consists of a flow of electrons, a copper wire, for instance, makes a good conductor of electricity. Paul Drude, in Germany, had become aware of this structure of metals by 1900.

Electrostatic bonds

Hydrogen bonds are a relatively weak and impermanent kind of electrostatic bond. Especially in colder water, a slightly positively charged part of one water molecule tends to stick to a slightly negatively charged part of another. In a sugar solution, for instance, one or more water molecules may become similarly attached to a sugar molecule. If one stands with bare feet on damp soil, there will be some hydrogen bonds between foot and soil. The two parallel portions of a double-strand DNA molecule are also held together by hydrogen bonds. And the adjacent folds of a single protein molecule may be similarly held together.

There is also ionic bonding. By 1887 Arrhenius in Sweden had realised that in a solution of common salt, NaCl, the so-called molecule splits up into two charged ions, Na^+ and Cl^-. But what is the situation in a solid crystal of common salt? When W. L. Bragg in 1912 studied the structure of a salt crystal, he found that the Na^+ and CL^- ions were stacked with beautiful regularity, but that each Na^+ was equally bonded to *six* Cl^- ions which were at equal distances from it. There were no pairs corresponding to the idea of an NaCl molecule. This 'upset the apple-cart'. W. L. Bragg was urged to look again. But such pairs are just not to be found. The same applies to a number of other substances. Consequently the term 'formula weight' rather than 'molecular weight' can be applied to substances such as common salt.

Covalent bonds

This third type of bond has been described above, with regard to H_2 and H_2O. Pauling notes that Gilbert Lewis himself evidently considered that only when a group of atoms was held together by covalent bonds, did the group constitute a proper molecule. Similarly Jacques Monod in *Chance and Necessity* (1972, p. 59) – written while he was Director of the Pasteur Institute in Paris – considers that only covalent bonds should be regarded as 'chemical bonds' in the strict sense of the term. And he further points out that in aqueous solutions covalent bonds are commonly several times stronger than non-covalent bonds.

Possibly there is a case for recognising, as a distinct category, those discrete groups of two or more atoms (either electrically neutral or charged) where each atom is joined to another by a covalent, or partially covalent, bond (Pettersson, 1979). Such a category would for instance include the sulphate ion, SO_4^{2-}, besides molecules such as those of water and glucose.

The water molecule, its sizes and mass

In crystals, including ice crystals, the arrangement and the distance apart of the different atomic nuclei can be worked out very accurately. The positions of the various electron orbits are relatively vague, but from the distances the nuclei are apart one can deduce atomic radii and knowing a radius, one can at least draw a circle, around the nucleus, to gain a rough idea of the outward shape and of the size of some molecules. As an example Fig. 4.4 is included. In Figs. 4.4 (b), (c) and (d) the single line between the atoms is a conventional

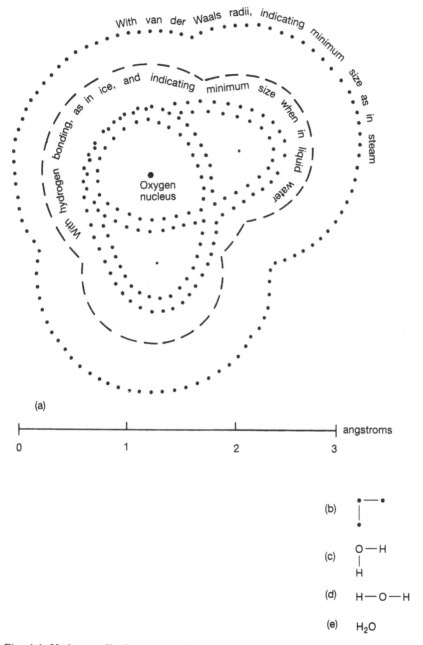

Fig. 4.4. Various stylised representations of one water molecule. (a) is a tentative scale drawing (×400 million), but with the bonding orbits purely schematic.

Complexity and evolution

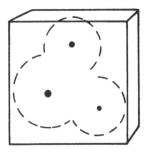

Fig. 4.5. Sketch of a water molecule, momentarily occupying an 'occupation volume' of 30 cubic angstroms. The side of the cube represents 3.1 angstroms, at a linear magnification of ×100 million. (Thus one angstrom becomes one centimetre.)

way of representing a single covalent bond, where *two* electrons orbit around the two atoms.

The actual dimensions of a water molecule are rather interesting for a further reason. It can help us to gain some idea of how close the molecules are in ordinary liquid water. It so happens that, since the weight of one cubic centimetre of *water* was taken to define the metric unit, one gram, it follows from the weight of the water molecule being 30 yoctograms that the average volume (in liquid water) occupied by each water molecule is about 30 cubic angstroms.

One can think of this small volume arranged in the shape of a cube, in which case the side of the cube would be 3.1 angstroms. Or one can think of it as a cuboid of which the sides are 4, 3 and 2.5 angstroms long. Looking now at the scale of Fig. 4.4, it is clear that the molecule very nearly fills its average occupation volume of 30 cubic angstroms (Fig. 4.5). In liquids and gases the molecules are in perpetual motion, barging into each other and bouncing apart. But these molecules of liquid water will be able to move only *very* short distances before bumping, and of course some will frequently be loosely linked together, for periods, by the weak electrostatic bonds called hydrogen bonds.

Hierarchical diagrams

The carbon atom and its parts

The carbon atom is a convenient Level 2 entity to analyse and plot, as in Fig. 4.6 At (a) we see the parts of the whole represented by the conventional five or so dots, standing for two, several or many parts. Diagram (b) illustrates the idea of 6 orbital electrons, and a nucleus composed of 6 protons and 6

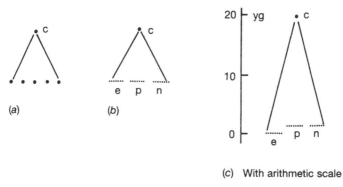

(c) With arithmetic scale

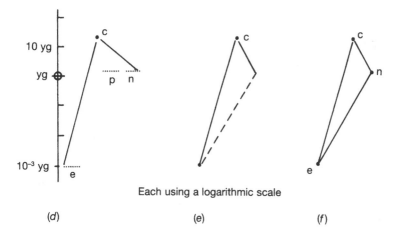

Each using a logarithmic scale

(d)   (e)   (f)

Fig. 4.6. Several of the ways of representing a carbon atom, as composed of fundamental particles. yg, (i.e. 10^{-24} gram). There is no quantitative horizontal scale.

neutrons. Then (c) shows the fundamental particles and the whole carbon atom, plotted according to their mass, along an arithmetic scale. And (d) shows the same for a logarithmic scale, with the mass of each electron as about a thousandth of a yoctogram.

In (e), while the maximum and minimum value for the mass of the fundamental particles is still indicated, the two points are joined by a broken line. A broken line is used rather than a dotted line, since the latter might give the impression that there was a continuous series of mass values between the minimum and the maximum, which as (d) shows is *not* the case. However, in (f) the minimum and maximum values for the parts of the whole are simply

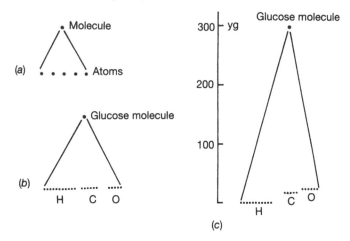

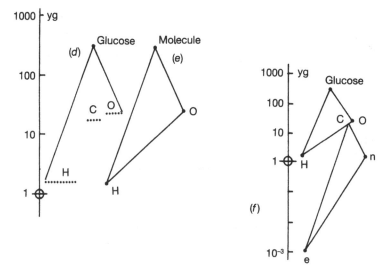

Fig. 4.7. Various ways of depicting the composition of a glucose molecule ($C_6H_{12}O_6$). (c) uses an arithmetic scale, and (d) to (f) a log scale.

joined by a continuous line. This diagram is a special gravimetric triangle, showing the minimum and maximum weight of the various parts, together with the mass of the whole of the parts.

The glucose molecule and its parts

The glucose molecule consists of 6 carbon, 12 hydrogen, and 6 oxygen atoms. Its mass is roughly 300 yoctograms, 10 times that of a water molecule. Diagrams (a) to (e), in Fig. 4.7 illustrate the glucose molecule in the same manner as the carbon atom in Fig. 4.6. Fig. 4.7(f) effects a link-up. Along a logarithmic scale of mass, a glucose molecule of about 300 yoctograms is shown as composed of its atoms, ranging from that of hydrogen to that of oxygen. And then one of the carbon atoms, at 20 yoctograms, is shown as composed of its fundamental particles, ranging in mass from 0.001 to 1.67 yoctograms.

Later we will be able to display a series of such quantitative gravimetric triangles, ranging from fundamental particle up to the society of a sovereign state.

Chapter 5
Biological range of integrated natural entities (first part)

Introduction

The biological range includes all individual living organisms, as well as many of their small and very small parts. Fig. 1.2 shows the first six levels of the natural hierarchy.

We must consider each of the three levels in detail. In the previous chapter some attention was paid to the structure and functioning of the physical range. Similarly it seems appropriate now to offer a very brief account of the structure and behaviour of organisms, cells, chromosomes and genes. It will also be interesting to enquire into the evolutionary origin of the organisation seen within ordinary cells. And we shall briefly review the range of mass of the members of each of these levels.

Level 6, multicellular organisms

When Robert Hooke in London published magnified pictures of sections through cork in 1665, he wrote of each tiny unit as a *cellula* (a little cell). That is how the biological term cell originated. Fig. 5.1 shows some cells, each with a nucleus, as seen in a section of tissue through the tip of a plant stem.

Aided by the new achromatic compound microscope of the 1830s, Schleiden and Schwann in Germany had by 1840 proposed that all larger plants and animals were composed of many tiny cells. Plants commonly have thick hard inert walls of cellulose around each cell (and running outside each cell's thin semi-permeable plasma membrane). Animal cells, however, are often separated only by their thin plasma membranes, so it took zoologists longer to become convinced of the cell theory. By 1900 however it was generally agreed that larger animals too were 'multicellular' organisms. In recent years

Tip of growing stem

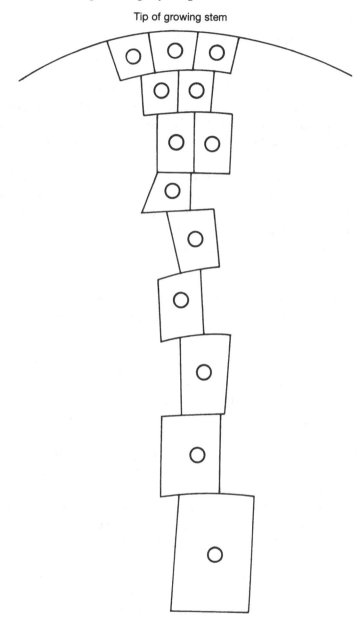

Fig. 5.1. Sketch of some plant cells, as in the growing point of a plant stem. Each is demarcated by a cell wall and each has one nucleus.

it has become clear that it is the nucleus of each ordinary cell which acts as its biochemical headquarters, from which chemical messages emanate.

Some features of living organisms

Each individual living organism – from the size of a bacterium upwards – has a tendency to increase in size (to grow). It commonly produces or helps to produce new small individuals like itself (it reproduces). And it will eventually die.

A multicellular organism commonly starts as a single cell. There is repeated cell division and the cells differentiate. Often the original single cell is formed by the fusion of two cells (sex cells or gametes) which usually originate from different adult individuals.

Each individual living organism contains nucleic acid and protein. The subcellular organisms called viruses – which are on the borderline between animate and inanimate – also contain nucleic acid and protein. If the individual is as large as a bacterial cell, it will also contain a range of carbohydrates and fats etc, besides mineral salts and usually a good deal of water. It contains many enzymes, being those proteins which promote specific chemical reactions.

It may be mentioned that nucleic acids are of two general kinds, RNA and DNA. Each molecule contains many sugar units. In RNA or ribonucleic acid, the sugar used is ribose. But there is another sugar, deoxyribose, which contains one less oxygen atom (hence its name). And it is that sugar which occurs in deoxyribonucleic acid or DNA.

If an organism is as large as a bacterial cell, the individual will take in food molecules and eliminate waste material. Many kinds of organism need to take in quite a range of complex molecules; and animals take them in in bulk, through the mouth. But green plants, for instance, need only simpler molecules. Using energy from the sun and aided by the green substance called chlorophyll, carbon dioxide and water are used to make up sugars, by the process known as photosynthesis. These sugars and similar molecules form the essential raw material from which all other compounds, in plants and animals, are eventually made. Oxygen is a waste product from photosynthesis. Most cells however consume oxygen so as to release, from larger molecules, the energy which powers the other living processes. The taking in of oxygen and giving out of carbon dioxide can be referred to as respiration.

Living organisms and their cells are able to react, to respond, to various stimuli. Animals have specially developed nerve cells. These permit communication between different parts of the body, and also the precise appreciation of (for instance) touch, taste, light and sound. All but the smaller animals have a marvellous compact assemblage of many nerve cells which we call the brain. Green plants and fungi have nothing comparable. There are many other kinds of specialised cells, for instance, the elongated cells of striated muscle used in voluntary actions where each cell has many nuclei. Plants do of course also respond to various stimuli, for instance by bending movements. Leaves sometimes move to secure better light, and a daisy partially closes up in the evening.

Level 5, ordinary cells with nuclei

An enormous number of ordinary cells have just a single nucleus. But some, like the cells of striated muscle, have a number of nuclei. On the other hand, some cells lose their nuclei before their main function within the organism commences. A familiar instance is that of our red blood cells which travel round our body delivering oxygen. Each had a nucleus at first, but it is lost as the cell matures and the marathon journeys commence.

A more extreme case is that of the elongated tubular 'vessel elements' in the veins of plants, which help to conduct water upwards. These have not only lost their nuclei, but all the rest of their protoplasm as well. They have lost the whole of their 'protoplast', as indeed the cork cells observed by Robert Hooke would also have lost all their contents.

On learning about plant cells, it may be thought at first that the protoplast of each cell is completely separated from that of neighbouring cells. But this is not quite correct. It has been known for many years that some thin strands of protoplasm (the plasmodesmata) do pass through the cell walls, and so join up protoplast to protoplast. The strands are often 30 nanometres or more wide, and provide a path for entities as big as viruses to pass from one cell to the next. This has been photographed with the electron microscope. Indeed the plasma membrane around one protoplast passes through such connections and is thus continuous with the plasma membrane of the next cell, and the next cell, and so on. Somewhat thinner connections also occur between many animal cells.

Level 4, intermediate entities

Each chromosome-plus-products unit is thought of as a member of Level 4. This idea of low-grade biological units, each centred upon one chromosome, is still relatively new and not widely known. The idea of intermediate entities has been described in the scientific literature as long ago as the 1970s, (for instance in Pettersson 1976, 1978*b*, 1979). Certainly 'intermediate entities' do seem very appropriate members of an integrative level which occupies a position between that of molecules and that of ordinary cells with nuclei.

In an ordinary cell the products of one chromosome will mingle and react with the products of other chromosomes. Level 2, that of atoms, provides a helpful analogy. Here some electrons are known to orbit around not one but two atomic nuclei. And in metals, some completely detached electrons wander widely; or flow along as an electric current. However one still confidently assesses the number of *atoms* present by counting the number of atomic nuclei, just as in a biological system one can similarly assess the number of *intermediate entities* present, by counting the number of chromosomes.

A little more will be said about intermediate entities in the next chapter. For the moment we will discuss the behaviour of ordinary cells, chromosomes and genes. Table 2.2 reminds us of the six kinds of integrated natural entity which are members of the physical and biological ranges.

Cell division, fertilisation and genes

Many multicellular individuals each start as a single cell, as a fertilised egg. The smaller male gamete, the sperm cell, swims up to the larger female gamete, or egg cell. The nucleus of the male gamete enters the egg cell and fuses with the female nucleus, to produce the nucleus of a fertilised egg. Each gamete had the half (or haploid) number of chromosomes, while the fertilised egg of course has the double (or diploid) number – such as will normally be present in the other cells of the adult multicellular organism.

In humans and other mammals, the female gamete or egg contains in its nucleus a so-called X chromosome. If the male gamete also contains an X chromosome, the new individual will be XX and female. But if the male gamete happens to contain a Y chromosome instead, an XY or male individual is produced. In humans, the egg cell is nearly a tenth of a millimetre in diameter; and the full diploid number of chromosomes is 46.

Biological nuclei are usually spherical, bounded by a membrane, and commonly several micrometres in diameter; usually less than 10 micrometres. It is often not realised that each chromosome – when not coiled up – may be a millimetre or more in length. That is the condition when it can be chemically most active, as in a cell which is not in the process of dividing. A useful analogy (for the chromosomes in a nucleus) is to think of a goldfish bowl some 20 centimetres in diameter, filled with water and, floating about

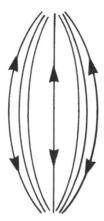

Fig. 5.2. During nuclear division a spindle forms and the chromosomes move from its middle, the equator, towards the two opposite poles. For clarity, only three chromosomes of each group are shown.

in it, several dozen thin threads each around 50 metres long. This extended condition takes the individual chromosome beyond the resolution of a light microscope, but they can, however, be photographed using an electron microscope.

Before a cell divides, to make two cells, each chromosome has to become two similar chromosomes. This occurs by the two parallel strips of DNA in the chromosome first coming apart. On each of these single strips an appropriate partner strip then forms; so that (when some further protein becomes attached) two complete chromosomes will now be present. These then coil, and coil again, to get down to a manageable length. They go to the middle ('equator') of a structure known as the spindle, from which they move apart, as though pulled, to the respective poles (Fig. 5.2). Of each newly formed pair of chromosomes, one goes to one pole and one to the other. Each new complete group becomes surrounded by a nuclear membrane; and the cell itself then divides into two cells.

In their coiled condition the chromosomes are of course much thicker, and *can* be seen with the light microscope. They take up various stains. Hence the Greek-derived name chromosome, for 'coloured body'. This normal kind of very frequent nuclear division is called mitosis.

When gametes are formed by a diploid parent 'meiosis' takes place. The cell of a diploid adult has two of each kind of chromosome, one having originally come from the father and one from the mother. Each gamete will get only one of each pair of similar chromosomes.

Crossing over

Without detailing all the stages of this kind of cell division, we may briefly note one curious complication known as 'crossing over'. After the two chromosomes of a pair have come together, a kink commonly develops, at one or occasionally two places along the length, and *new* chromosomes are formed containing portions of each of the two original chromosomes. Crossing over is illustrated in Fig. 5.3. In the first of the three diagrams, (*a*), two similar chromosomes are shown, one having come from the father and the other from the mother of the individual. They are shown as differing in various comparable (lettered) genes. For instance one gene of a pair might lead to coloured flowers and the other to white flowers; or, in an insect, one might cause red eyes as opposed to normal eyes.

The two similar chromosomes come together, (*b*). But there is commonly a twist or kink. A 'cross-over' occurs. They then separate as in (*c*) and the two somewhat modified chromosomes go their ways to the different gametes.

This odd behaviour happens to have been very helpful to investigators. In the early years of the century, breeding experiments with many species served to illustrate and confirm the famous laws of hereditary initiated by Mendel. But there was no information about *where*, in a cell, the 'factors' for this or that character were situated. And it was due to the phenomenon of crossing over that it became clear that the factors, or 'genes', were in fact situated upon the chromosome of a cell.

Using the present knowledge that genes are situated on, or indeed in, chromosomes, Fig. 5.3 can help to make the proof clear. Supposing a crossover to occur more or less anywhere along the pair of chromosomes, at random, genes B and C will very seldom become separated, being so close. B and D (at twice the distance) would become separated about twice as often, though still not very often. If the adult has many offspring, the characters corresponding to genes B, C and D will all tend to be inherited in a linked manner. They are seen to belong to the same 'linkage group'. Genes A and M will usually get separated. Genes G and M, for instance, may get separated about half of the time. But if one is able to do breeding experiments with a dozen or so genes (A to M), all spaced out along the same chromosome, it will be found that A is closely linked to B, B is closely linked to C, C to D, D to E, and so on, to L to M. It becomes clear that all belong (directly or indirectly) to the same linkage group.

Quite early, research workers in America concentrated on a very small fly, the fruit-fly, the common species of *Drosophila*. In one's home one sometimes sees it hovering around very ripe bananas. A family of over a hundred

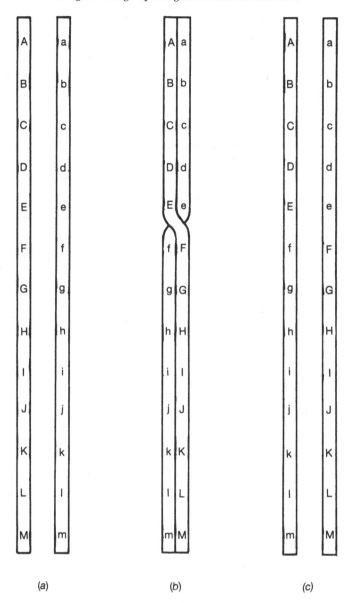

(a)   (b)   (c)

Fig. 5.3. Diagrams illustrating crossing over. For explanation please see text.

flies can be raised in a vessel no larger than a 250 millilitre bottle, in a fortnight or less. The fly has only four pairs of chromosomes per cell; and the many characters studied were all found to be grouped into just four linkage groups. *Therefore*, it was correctly concluded, the factors or genes are very probably situated on the chromosome. T. H. Morgan had reached this explanation of linkage groups by 1911.

The point was made certain by the discovery that the same important match between (a) the number of linkage groups, and (b) the number of pairs of chromosomes per cell, occurs also in other species, whenever they are exhaustively investigated. Thus some other related species of *Drosophila* happen to have three, five or six pairs of chromosomes, and the corresponding number of linkage groups. The number of linkage groups and of pairs of chromosomes in maize, for instance, is ten; and in green peas seven. In some research departments, Morgan's idea was not accepted for many years after his paper of 1911. The reason is as follows.

Drosophila is small. It is cheap to breed and it has a very short generation time. If working with sweet peas or rabbits, for instance, the results come in much more slowly. What happened was that – with reference to Fig. 5.3 – genes A and B, for instance, were found to be associated; and while gene C did *not* seem linked to A and B, it *did* seem linked to H. So these genes were reported as belonging to two different linkage groups. All in all, a distinctly greater number of linkage groups was reported, than the number of pairs of chromosomes. Hence such experiments suggested that the factors or genes were *not* situated upon the chromosomes. In retrospect, one sees that these experiments were not sufficiently complete and exhaustive. Another bonus from linkage studies is being able to map out the relative position of different genes, along a chromosome.

Triplets and gene action

It was in 1953 that Watson and Crick, at Cambridge, discovered that a chromosome (apart from its protein) consists of a spiral of two long strands of DNA, running parallel along the spiral (Fig. 5.4). They are held together by hydrogen bonding. Each DNA strand consists of a series of nucleotides, each having about three dozen atoms. A nucleotide has three main parts: 'base' (so-called), sugar and phosphate (Fig. 5.5). In any DNA nucleotide its base is one of four kinds, either an adenine, a thymine, a guanine or a cytosine. Each base has a dozen or so atoms; and it is customary to refer to the bases by their initial letters, A, T, G and C.

Fig. 5.6 shows a very small portion of the two parallel strands of DNA of a chromosome. The nucleotides are stacked relatively flat, along the strands, and their average thickness is only about 3.4 angstroms, comparable to the size of a molecule of water (Fig. 4.4). To help the nucleotides fit snugly together (aided by hydrogen bonding) opposite each A there is a T, and opposite each G there is a C. On one side of Fig. 5.6 a group of three

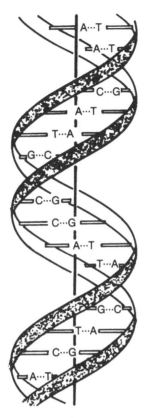

Fig. 5.4. A schematic diagram of a small portion of the DNA double helix. The dotted lines represent the hydrogen bonds between the two strands. The structure actually contains less space than the diagram indicates: it is really more like a solid rod with two grooves running round it. (From Watson 1970.)

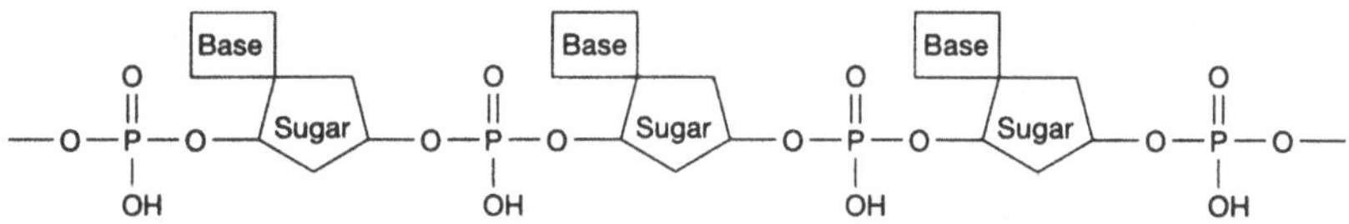

Fig. 5.5. A small section of a nucleic acid chain. Three-and-a-bit nucleotides are depicted, since one nucleotide consists of base plus sugar plus phosphate. (From Cohen 1965.)

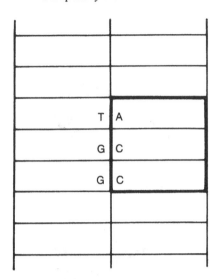

Fig. 5.6. A small portion of the two parallel strands of DNA in a chromosome, strands which are held together by hydrogen bonding. The triplet containing ACC is outlined (see text).

nucleotides is shown, with a heavier outline, being one 'triplet'. It is easy to see that there can be many such triplets, e.g. ATA, ATT, ATG ATC and so on.

Since the thickness of each nucleotide, stacked in a triplet, is about 3.4 angstroms, the triplet itself will be about 10.2 angstroms long, or just about one nanometre. And the average mass of a triplet, as bonded in a chain, is roughly 1500 yoctograms or 1.5×10^{-21} grams.

A gene with a thousand triplets would be about a thousand nanometres or one micrometre long; and a chromosome with a thousand such genes would be about a millimetre long (Fig. 5.7). Correspondingly, the mass of such a gene (represented by a single strand of DNA) would be around 1.5×10^{-18} grams; and the mass of the single strand of DNA running along the whole millimetre chromosome would be around 1.5×10^{-15} grams. The longest chromosome in the human nucleus is estimated to be some 8.5 centimetres long, while the weight of each of its DNA strands is apparently about 1.3×10^{-13} grams.

To understand how a gene, in a nucleus, causes a certain protein molecule to be made, outside the nucleus, think of a portion of a chromosome, a certain gene, which contains a thousand triplets, and which is thus about one micrometre long (Fig. 5.7). Alongside this strip of DNA, a long complementary strip of the similar substance RNA is formed. The original strip acts as a

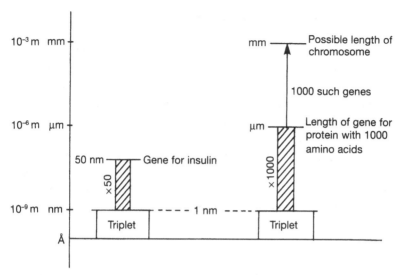

Fig. 5.7. Logarithmic plot of some *approximate* lengths along the DNA of a chromosome. The level of one angstrom seems to form a convenient baseline for this particular kind of diagram.

'template'. With regard to the bases it contains, RNA is similar to DNA except that the base thymine is replaced by the base uracil and, as mentioned earlier, RNA and DNA also contain different sugar units.

The new strip of RNA becomes detached, and passes out of the nucleus into the surrounding cytoplasm, where it causes the making of a certain kind of protein molecule. A specialised particle called a ribosome (some 20 nanometres in size) moves along the strip of RNA, and for every triplet it comes to, the ribosome catches the appropriate amino acid, from a stock nearby, and step by step builds up the protein chain of a thousand amino acid units. Two specific examples of the 'genetic code' are (1) the triplet ACC in a gene causes the amino acid tryptophan to be added by the ribosome, and (2) the triplet AAG causes phenylalanine to be added. The same strip of RNA can be used for making a succession of the same kind of protein molecule. Some proteins of course are enzymes, while others are units for structural building.

The foregoing is an outline of how a gene acts to cause a protein molecule to be made. Various complications of the process have been omitted. There are many questions still to be answered; but each decade new research deepens our understanding of this fascinating process.

The genetic code was worked out in detail in a number of laboratories, mainly in the 1960s. Each of the 20 amino acids found commonly in natural proteins is coded for by more than one of the different triplets. And the code is said to be 'universal', in that it commonly occurs in all living organisms from bacteria to humans.

The above example spoke of a gene containing a thousand triplets, and thus being about a micrometre long. (However both within and between genes there is sometimes some extra and inert material.) Fig. 5.7 shows how a chromosome with a thousand such genes would be about a millimetre long. And when investigations are made, it is often found that particular chromosomes are around one or more millimetres long. Protein molecules and the genes which code for them do however vary a good deal in size.

The gene which codes for the enzyme called β-galactosidase, for instance, is apparently 1.26 micrometres long. But the small protein insulin has only 51 amino acids; and so its necessary gene length would be little more than 50 nanometres.

Mutations

When a gene is copied to make two similar genes, sometimes – very rarely – a mistake occurs. A change in one triplet will usually make a change of one amino acid in the protein produced. One speaks of a mutation having occurred; and sometimes a new individual containing the different gene will obviously look or behave differently. Gene mutations can also be caused by damage from X-rays and ultraviolet light.

The great majority of gene mutations will be disadvantageous. But a very small proportion may be helpful, perhaps helpful in coping with life in some slightly changed environment. And it appears that such random changes have provided – accidentally as it were – the important new genes and groups of genes which have permitted the evolution of one species from another. It is said that variation plus natural selection gives evolution. And it is the mutation of genes (plus some other alterations of chromosomes) which ultimately causes variation among related individuals.

Chapter 6

Biological range of integrated natural entities (second part)

Levels 6, 5 and 4

It will now be clear that a chromosome is a pre-eminent biochemical head-quarters, which initiates the making of many special protein molecules. A biological nucleus, containing a number of chromosomes, is a higher-level biochemical headquarters; and a multicellular organism is a biological system at an even higher level.

We may briefly check that the two criteria of Chapter 2, the compositional and the duality criteria, are satisfied. Certainly each intermediate entity, at Level 4, is composed mainly of molecules which are themselves members of Level 3. Each ordinary cell, of Level 5, is composed mainly of intermediate entities of Level 4, and each multicellular organism at Level 6 is also composed (materially) mainly of Level 5 cells.

Some bacterial cells, those with a single chromosome, are themselves intermediate entities which live freely and independently. While many other intermediate entities collaborate in the constitution of ordinary cells. Some ordinary cells, such as amoeba and *Chlamydomonas* (Fig. 2.1) live independently, while many other cells collaborate in the constitution of multicellular organisms. And while most multicellular organisms live independently, during the past 200 million years the mammals, birds and some insects have come to live in social family groups, where the adult actively aids the young after hatching or birth. Thus all three of the biological levels can be seen to satisfy both the compositional and the duality criteria.

Bacterial cells

The cells of bacteria (in the narrow sense) are spherical or rod-shaped, and from less than a micrometre to several micrometres in size (Fig. 3.1).

53

However the cells of the Cyanobacteria (also called blue-green-algae) are often several times larger. Bacterial cells may separate after cell division; or may form chains which can be regarded as colonies of a number of individuals. Each bacterial cell contains one good-sized chromosome. But it is circular rather than linear.

The cell of an ordinary bacterium may have only a single chromosome. In that case it ranks as a single intermediate entity, a member of Level 4. But just before cell division there will be two chromosomes, so that two intermediate entities are then present; and occasionally there are four before the cell finally divides. In the centre of the cell there is a large circular chromosome. But not infrequently one or several extra small circular or linear chromosomes are also present in a bacterial cell. The special viruses (bacteriophages) which can enter and flourish within a bacterium normally have small linear chromosomes. However, plasmids normally have a small circular chromosome. Some plasmids (in the broad sense) may contain a gene for resistance to a medical drug. And a plasmid, or part of it, may sometimes join up as part of the main chromosome. Thus a single bacterial cell may comprise quite a number of intermediate entities.

When one looks at one bacterial cell, under a microscope, one can be sure of seeing a single cell, but one *cannot* tell how many intermediate entities are present. The same kind of numerical uncertainty can occur at the level of multicellular organisms. If one points to a hedgehog walking across the pavement and asks a naturalist how many multicellular organisms there are there, the answer will probably be a cautious one. For a number of small multicellular organisms known as fleas commonly lurk among the hard spines of the hedgehog and it might also have other multicellular parasites. While a bacterium, just before cell division, will contain more than one main chromosome, so also a pregnant hedgehog will also contain further multicellular organisms of the same species.

The case of a hedgehog is not so exceptional. Under wild conditions, it may well be that more than half the individual mammals likewise harbour small multicellular organisms in the form of fleas, lice, ticks and several kinds of worms. Even whales have barnacles and lice, apart from internal parasites. Gerald Durrell (1964) mentions the ever present task of de-fleaing, de-lousing and de-worming captured animals, prior to taking them to zoos.

Viruses

The many different kinds of virus cause many diseases. These include influenza and poliomyelitis in humans, foot and mouth disease in cattle, and the various mosaic diseases in plants. In all these instances the viruses flourish and multiply within the cells of the diseases organism. And a special group of viruses, the bacteriophages, can enter and multiply in bacterial cells.

If the sap from a diseased tobacco leaf, after it has been passed through a porcelain filter which removes all bacteria, can *still* infect another tobacco leaf with the mosaic disease, evidently some entities still smaller than bacteria are involved. That viruses are indeed smaller than bacteria was already indicated in Fig. 3.1. And since they reproduce in the host cells, one naturally tends to think of them as alive.

Chloroplasts and mitochondria

In many of the cells of green plants there are small green bodies which are responsible for the overall coloration. These chloroplasts contain the substance chlorophyll, which is the necessary agent for photosynthesis. A simplified version of the chemical equation is given below.

$$6CO_2 + 6H_2O + \text{light energy} = C_6H_{12}O_6 + 6O_2$$

Various kinds of sugar and other molecules are produced, and these are the raw materials from which all the other organic molecules of living organisms are eventually made. Chloroplasts are commonly discus-shaped and sometimes about 3 micrometres in diameter. Their importance in making carbohydrates has been recognised since the 1880s.

The ordinary cells of both plants and animals also contain mitochondria. Each mitochondrion is a colourless sausage-shaped body, commonly several micrometres in length. Since the 1940s it has been realised that mitochondria are concerned with practically all the enzymes involved in respiration. Sugar and oxygen enter a mitochondrion, carbon dioxide and water go out as waste products, and the energy-rich molecules of adenosine triphosphate (ATP) go out from the mitochondrion to power all the other living processes of the cell. A number of the distinctive proteins used by mitochondria are in fact coded for by genes within the cell's nucleus.

In the history of the study of disease, it is convenient to think of bacteria becoming recognised and established as a coherent group between 1840 and 1880, while viruses became recognised and established as a group between 1880 and 1920, roughly 40 years later. In the 1930s viruses were thought of as small living organisms. Some specialists take a different view today, since viruses (a) are not cells, and (b) have no respiratory enzymes. On the other hand it is convenient to regard viruses as *subcellular* organisms, a status which is also sometimes accorded to plasmids.

Each virus, in its complete condition, consists of one (or occasionally more than one) chromosome, with double- or single-strand DNA or RNA, surrounded by many copies of one or more kind of protein molecule. All is held together by hydrogen bonding.

With regard to our natural hierarchy, each virus particle (if with no more than one chromosome), or each similar virus-plus-products unit, is clearly an intermediate entity. Most other intermediate entities are also subcellular; and in an ordinary cell hardly any intermediate entities except those associated with mitochondria have respiratory enzymes.

The origin of chloroplasts and mitochondria

Chloroplasts and mitochondria each increase in number by growing larger and then dividing into two in the manner of bacterial cells. Each contains a number of copies of its small circular chromosome. And since around 1970 specialists have held the view that each mitochondrion is a descendant of a bacterium-type cell which long ago was taken into a larger cell and then kept there as a welcome guest because of the efficient way it used food to supply energy through oxygen-using (or aerobic) respiration. Similarly each chloroplast is thought of as the descendant of a green bacterium-type cell which entered a larger cell and was kept as a good source of carbohydrate food (Fig. 6.1). This process could be termed mutualistic symbiosis. Several authors proposed this idea in a special number of *American Naturalist* in 1965. It was certainly the dominent view at a meeting of the Royal Society in April 1978 (Richmond & Smith 1979).

Although now they each have more than one copy of their chromosome, one may think of chloroplasts and mitochondria as representing bacterium-like ancestors each with presumably just a single circular chromosome. The cell has (one might say) a number of small individual organisms living within it.

It has been mentioned that chloroplasts and mitochondria have rather short chromosomes, although a number of copies of each. It appears that in the long process of mutual adaption, many of their chemical needs have come to be served by the chromosomes in the nucleus, so that not so many different genes are now necessary in those small circular chromosomes.

The customary discoid chloroplasts of higher plants may have had a single origin. But the process of taking in and keeping a photosynthetic intermediate entity seems also to have occurred in at least two other lineages. Two different kinds of bacteria have provided their descendants within the cells of the brown algae and the red algae. And similarly a second intermediate entity, good at aerobic respiration, has been taken in and adopted on at least one

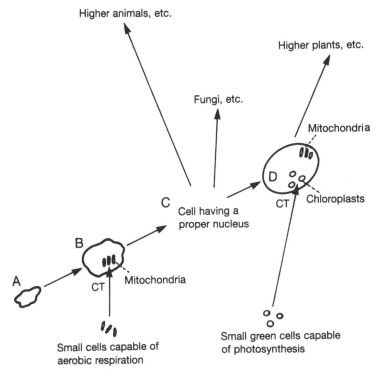

Fig. 6.1. Diagram of the supposed course of evolution of several main types of cell. Cell C represents the common ancestor of all organisms having a proper nucleus and the two customary forms of nuclear division. CT, 'coming together'.

other occasion, since one species of amoeba contains an aerobic bacterium which differs from the usual mitochondrion.

Ranges of mass

Having carefully defined membership of the three major integrative levels in the biological range, in addition to the three in the physical range, it is interesting to see how the ranges of mass values of the entities in each of the six categories compare (Fig. 6.2). Most of the integrated natural entities have certainly been included, but a few outlying ones may have been omitted. For instance, how should we estimate the mass of the very heaviest molecules? The largest diamond weighs well over 100 grams. Some quartz molecules are heavier still; and the heaviest mineral molecules will no doubt weigh many kilograms.

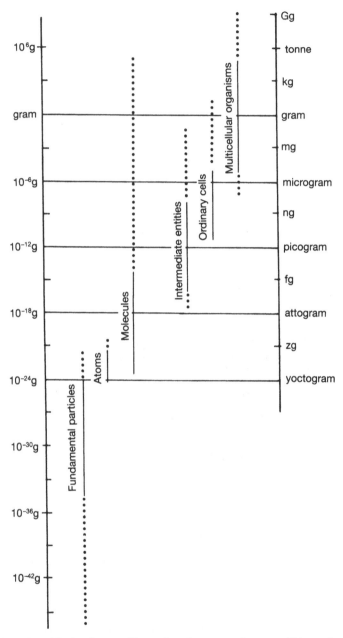

Fig. 6.2. A logarithmic diagram illustrating the range of mass within each of the major Levels in the physical range and in the biological range. A continuous line indicates, in an approximate manner, where the greatest concentration of entities appears to occur; and the rest are mainly covered by the dotted extensions. (But please see text.) About photons: (a) when moving they do of course possess mass; and (b) here only the photons within the solar system proper are considered, and not all those in the 'extended solar system' referred to in Chapter 12.

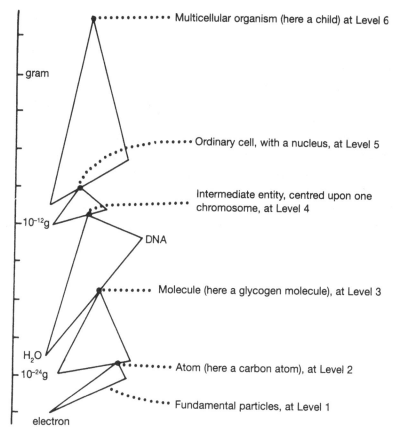

Fig. 6.3. A quantitative model of the natural hierarchy, up to the level of multicellular organism. The apex of each triangle is plotted according to the mass of a certain entity, and the two lower corners represent its lightest and heaviest components of the level below. There is no horizontal scale.

Six-level hierarchy of triangles

Using the logarithmic scale of mass, the material composition of one person (in this case a child) can be shown quantitatively by the top triangle of Fig. 6.3. The maximum and minimum mass of the integrated natural entities of Level 5 and Level 4, can only be approximately estimated.

Let us turn to the range of mass of the many kinds of molecules in the human body. While many human chromosomes are potentially several centimetres long, the longest chromosome is some 8.5 centimetres. The mass of the DNA double spiral of this chromosome is estimated at about 2.6×10^{-13} grams. That provides an estimate of the maximum mass of any molecule within the human body. (Alternatively, one can consider the double spiral as consisting of *two* molecules.)

The minimum end of the molecular range is occupied by our abundant water molecule, H_2O. This has a mass of about 30 yoctograms, or 30×10^{-24} grams. Hence the heaviest molecule in our bodies is at least eight thousand million times the mass of the lightest. This relationship is represented in Fig. 6.3 by the line from H_2O to DNA.

In Fig. 6.3 a large carbohydrate molecule of glycogen, or 'animal starch', has been chosen for further analysis into its constituent atoms of hydrogen, carbon and oxygen. Taking a medium-sized glycogen molecule, this has been plotted as weighing 10^{-17} grams. Below, just one of the carbon atoms, in such a glycogen molecule, is shown as anlysed into its range of fundamental particles, of which of course electrons are the lightest and neutrons the heaviest.

Chapter 7
Social range of integrated natural entities

Introduction

We have discussed in some detail the integrated natural entities of Levels 1, 2 and 3, in the physical range; and those of Levels 4, 5 and 6 in the biological range. So we now come to the social range: there are the one-mother family societies at Level 7, the multifamily societies at Level 8, and the societies of sovereign states at Level 9 (Fig. 7.1).

The zoological species with which this chapter is primarily concerned are those where the adult *actively* aids the young after hatching or birth. These species can be regarded as truly social species. However, the words social and society can be used with many different meanings. Some biologists may regard a group of plants growing together, and of the same species as constituting a society. The word actively, as used above, implies the use of respirational energy.

According to the evidence of fossils and isotopic dating, the first animals where the adult actively aided the young, after hatching or birth, were the early mammals. These lived about 200 million years ago. Similar societies of birds appeared about 150 million years ago. And the earliest social insect, found only recently, dates from about 100 million years ago.

There have been fascinating recent studies of the social behaviour of, for example, chimpanzees, orang-utans, baboons and other primates, of lions, tigers, wild dogs, seals, elephants, herring gulls, jackdaws and rooks, of certain fish and of bees, ants and termites. Even in the common earwig (*Forficula* in the Dermaptera) the mother guards the eggs and continues to protect the tiny earwigs for a short period after they hatch. So the earwig represents a still further insect lineage which has evolved to become social.

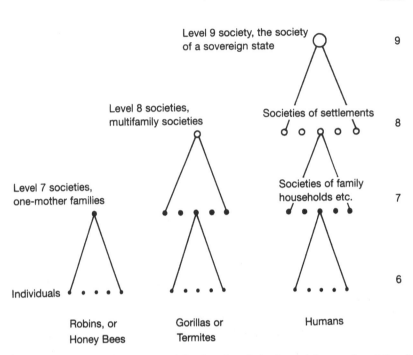

Fig. 7.1. Examples and structures of the three Levels in the social range. Level 7 and Level 8 societies occur in many social species, but Level 9 societies in humans only.

Level 7: One-mother family societies

In a truly social zoological species, a unit family society is a coherent group of two or more individuals living together, a group which typically comprises one parent (normally the reproductive female or mother) or both parents, plus the eggs or young which are being cared for. Such groups may persist after the time when the young become autonomous; and in some cases the family society may be recognised as such during the period before the young are produced. Some further discussion of Level 7 entities among humans follows in Chapter 8.

In the various species of animal where the young are looked after by only one parent rather than by two, it is in fact the mother who normally performs this function. This item of behaviour has arisen independently in a number of different lineages. No one need be surprised about it being the mother rather than the father, since it is from the mother's and not the father's body that the bulky eggs or young finally emerge. What are remarkable are the few odd cases where it is only the reproductive male or father who looks

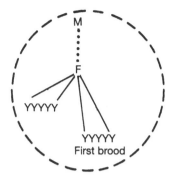

Fig. 7.2. Robin family.

after the eggs after they are laid. In the stickleback (a small freshwater fish) the father not only protects the eggs, but continues to look after the small fishes for a time after the eggs have hatched.

With regard to the theory of major integrative levels, the integrated natural entities of Level 7 conform to the compositional criterion, in that each unit family is an integrated group of Level 6 entities; and it conforms to the duality criterion in that some families, as in robins, exist as independent entities, while in other species, for instance the mountain gorilla, the unit families participate as components of multifamily societies of Level 8.

Robins

A robin family as studied by David Lack (1965) can be depicted diagrammatically as in Fig. 7.2. The M stands for the reproductive male bird and the F stands for the reproductive female or mother. The dotted line represents the passage of male cells. Two broods of young (Y) are shown, the second brood perhaps still being incubated by the mother, while the first brood is now being fed by the father. The unbroken lines from mother to young can symbolise that the young actually came from the body of the mother. This Level 7 society of robins is an independent society, and not fused into any larger social group. In fact the male bird will chase other robins off the family territory, and he continually proclaims the boundary of the territory by singing at the outposts. In the diagram, the broken line surrounding one family can be used to symbolise this independent status.

It was of course the European robin which Lack studied, not the American robin. But the one-mother families of many kinds of bird remain independent in much the same way. A colony of herring gulls, as studied by Tinbergen

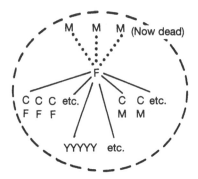

Fig. 7.3. Honey bee family. (See text.)

(1965), contains many families, but they are not integrated into multifamily society.

Honey bees

There are far more individuals in the family society of a beehive (Fig. 7.3), often tens of thousands, but there is only a single reproductive female so that it is still a Level 7 society. Though it is not always realised, the queen is often impregnated by several males, before commencing her marathon of egg laying. In the diagram CF stands for celibate female, that is for a worker who is in fact sterile, while CM stands for celibate male, for a drone who has not yet functioned. In the hive depicted, the young are now being cared for by their elder sisters (CFs), while the workers who were sisters of the present queen are now dead.

Level 7 societies sometimes contain more individuals than just the parents and young, though they are usually relatives. This is exemplified by the honey bees, where the celibate adult females remain in the group and perform most of the meticulous and painstaking work involved. Normally of course one can equate the number of Level 7 societies with the number of reproductive females.

During the few days of supersedure, an old queen bee with failing powers functions in the same hive as the young queen, her daughter, who is about to take over the whole family organisation. So for this short period the population of the hive can perhaps best be regarded as consisting of two Level 7 societies.

The bees of a hive keep themselves separate from other groups of bees. They constitute an independent society, like that of a robin family. But a modern human family – though it may occupy, when at home, a distinct

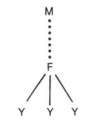

Fig. 7.4. Human family.

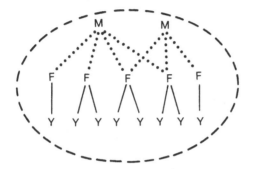

Fig. 7.5. Multifamily band of mountain gorillas. (See text.)

house and garden – is politically fused with others into a larger social group. It is not 'independent' in the manner of a robin family, and thus there is no broken line surrounding the human family in Fig. 7.4.

Perhaps a very general comment may be appropriate at this point. A mother, a chromosome and an atomic nucleus have somewhat similar roles, at three different integrative levels. Each Level 2 entity is centred upon a atomic nucleus; each Level 4 entity is centred upon a chromosome; and each normal Level 7 entity is centred upon a mother.

Level 8: Multifamily societies

Level 8 societies occur both in some vertebrate and in some invertebrate social species. Each multifamily society contains more than one reproductive female, and typically it is integrated into a coherent group with internal differentiation.

Mountain Gorillas

Our distant relatives the gorillas occur in tropical Africa, and the mountain gorilla is one of the subspecies (Fig. 7.5). *The Year of the Gorilla* (1965) by

George B. Schaller, from the University of Wisconsin, gives an excellent description of their way of life, especially in chapters 6, 7 and 8.

Schaller studied 10 of the independent wandering bands, which contained between 5 and 27 individuals. The young, up to three years old, can be ranked as infants. From three to six they are juveniles. The females then become sexually mature and soon reach their adult weight of nearly 90 kilograms. But from six to ten a male remains immature, as a 'blackbacked male'. After about 10 years he becomes a mature 'silverbacked male', and reaches about 170 kilograms in weight and a height of 1.7 metres. Gorillas may live for about 30 years.

In the 10 societies studied by Schaller, there were, on average, about 1.5 silverbacked males, 1.5 blackbacks, 6 females, 3 juveniles and 5 infants, making a total of 17 in all. A band obtained its plant foodstuffs while wandering over a 'home range' of some 30 square kilometres, and the home range of one band commonly overlapped with that of one or two others.

Where there is more than one silverbacked male in the band, there is no exclusive relationship between male and female. The different members of a band do of course recognise each other as individuals. They seem to lead relatively placid, contented, gentle lives; and to find contentment and satisfaction in being with and near the others with whom they are familiar. In contrast, they may become agitated if gorillas of another band are in the vicinity.

It is the leading male (not necessarily the heaviest) who decides (1) in which direction to travel during feeding, and (2) where and when to bed down for the night. Except for the youngest, each gorilla constructs a rather simple nest for sleeping in. An infant of over 18 months often makes a nest too, for practice. But at night each infant normally cuddles up with its mother in her nest.

In recent years the number of gorillas has been declining due to humans impinging on their areas. Dian Fossey, author of *Gorillas in the Mist* (1983), devoted much of her time and energy to the study of gorillas and to their protection from impinging humans. And as a consequence, it seems, she was murdered.

Some other multifamily societies

For a long period our early human ancestors probably lived in similar wandering bands, and also with a very low population density. But after the advent of crop-farming, the group became the population of a settled village, and later towns and cities emerged.

In the Black Hills of South Dakota there flourished a rodent called the black-tailed prairie dog. It was first studied in detail by John A. King in 1955.

For most of the year each Level 8 society, sometimes called a 'coterie', assiduously maintains its own territory which contains several dozen shared burrows, and the knowledge of the territorial boundaries is passed on as a cultural tradition. These multifamily societies contain on average about 1.5 reproductive males, 2.5 reproductive females and 6 young, making a total of about 10. But a large compact colony or 'town' of prairie dogs, perhaps a kilometre long, may contain a hundred or so Level 8 societies, each living within its own territorial limits. The Level 8 societies help each other by passing on an alarm call, if a predator such as a hawk approaches. But the group is not sufficiently integrated and internally differentiated to be classed as a Level 9 society.

Turning now to insects, the multifamily societies of various species of ants and termites are well known. As with honey bees, there are usually numerous sterile workers; and sometimes raids are made to bring in extra 'slave' workers. D. W. Morley (1953) mentions the large black ant, as being common throughout North America and temperate Eurasia. In a small colony with 200 to 300 workers there may be only two or three reproductive females or 'queens'. But there will be a large number of queens in a colony of 2000 to 3000 individuals.

Level 9 Societies of sovereign states

Such entities occur only in our own remarkable species *Homo sapiens*. Each is an integrated group of Level 8 societies, of the societies of settlements and each has at least moderately well-developed internal differentiation, for instance with some persons functioning as administrators.

The category clearly conforms to the compositional criterion of the theory of major integrative levels. But the duality criterion is not here relevant, since being at the highest level one cannot expect to find Level 9 entities which are constituent members of any *fully-integrated* entity of a level above. As will be discussed in Chapter 9, it may be around 8000 years ago that the first Level 9 entity appeared on Earth.

Characteristic features of truly social species and their societies

Professional research has been made into the general characteristics of unit family societies and of the species in which they occur. For instance what kinds of aid are given to the young? Which kind of aid (in evolution) was the first? And in what other ways do such species differ from non-social species? But no standard work or well-known paper has so far been found to supply general answers. So the next two paragraphs certainly provide a novel statement, even if they contain nothing unexpected. The *first* way of aiding the young seems normally to be by protection – rather than by feeding – to judge from the many species which give some protection to their young without feeding them; or those who protect their eggs but without protecting the hatched young.

Among truly social species, a typical family society is an integrated group of related, unequal but cooperating individuals of the same species. One or more of the adults actively aids the young after hatching or birth:

(1) By protection, for instance from predators, parasites, pathogens, molestation, and from a variety of adverse conditions
(2) Commonly by feeding
(3) Often by cleaning
(4) Often by the occupation of a family area (territory or home range)
(5) Sometimes by constructing 'an artifact for living in', a family home.

In truly social species there is a relatively low death rate among the young, and in birds and mammals the young produced are relatively few and large. Furthermore such species are characterised by relatively well-developed communication between individuals. It is a fascinating but little known fact that the evolution of a supply of 'milk' for the young has occurred several times quite independently: in bees, in pigeons, and twice it seems in mammals, for while the monotremes (such as the duck-billed platypus) make use of modified grease glands, most mammals use modified sweat glands for secreting milk. Doubtless further features characteristic of truly social aspects will come to be recognised. Do they, for instance, possess a higher proportion of nerve cells in comparison with other animals of the same size and shape? This seems likely to be the case among vertebrates, but is it so among insects?

Conclusion

Employing the compositional criterion and the duality criterion of the theory of major integrative levels, the series has now been extended in a precise manner up to Level 9. The 'social associations' of which Joseph Needham spoke at Oxford in 1937 have been regrouped into three major integrative levels.

The whole of social evolution seems to have been confined to about the last 200 million years. Earlier biological evolution had a duration much more than ten times as long. Fig. 2.4 includes some of the approximate dates, and serves as a reminder of the thick strata of lower-level complexities from which family societies eventually emerged.

Chapter 8
Human societies (first part)

Introduction

Three centuries ago Archbishop Ussher of Armagh, in Ireland, made a careful quantitative study of the evidence in the Old Testament of the Bible, and concluded that the first farming or agriculture (as well as the first humans) occurred about 6000 years ago. Nowadays, as a result of many careful excavations and with isotopic dating, it is concluded that farming or agriculture first appeared before 10 000 years ago, though it was roughly in the same part of the world as the Bible suggests.

Whether or not one invokes supernatural intervention to help account for subsequent changes, since the beginning of agriculture the ways of human life have come to diverge more and more away from those of our other mammalian cousins. For one thing there has been the increase in the number, diversity and bulk of artifacts (of made things). Another general feature has been the increasing speed and distance of the movement of people, goods and information. In William Shakespeare's play *A Midsummer Night's Dream*, written nearly four centuries ago, a fairy messenger named Puck promises to 'put a girdle round about the Earth in forty minutes'. In fact nowadays radio messages can reach the other side of the Earth in less than a second.

It so happens that I used to have experience of something like that, night after night. In the early part of World War II one of my army jobs, in England, was to take down news items, broadcast by radio in the morse code, from the Domei News Agency in Japan. The particular broadcasts were being beamed from Japan towards America. By a series of reflections between ground and ionosphere, some of the radiation continued beyond America and arrived in England. But from the aerial system being used in Japan, some electromagnetic waves also leaked out of the back and came to England direct across Asia and Europe. Having a shorter journey these arrived first.

The radiation which crossed the Pacific, the American continent and then also the Atlantic arrived in a small fraction of a second later. So the end of each dot or dash was followed by an 'echo', which made it difficult to read. This is mentioned as a case of circumglobal communication which happens to have been experienced personally. The herring gulls of a large colony may give mutual warning of an approaching predator, like the rodents in a prairie dog colony. But how vastly different have the artifacts and communication of *Homo sapiens* become, these last few thousand years!

In recent times, the difference between the way of life of humans and of other animals has increased to a quite amazing extent. Thus when examining contemporary human societies, we shall find it desirable to modify the concepts of Level 7 and Level 8 entities, so as to provide a closer match with the present human situation.

Level 7: Families and households

In a truly social species, where the adult actively aids the young after hatching or birth, the standard Level 7 entity is a unit family, a one-mother society, typically consisting of a reproductive female and one or more of her young, sometimes with the father as well and occasionally including other individuals, usually relatives. It has also been noted how the term might be extended, in some cases, to a period before the young appear or after they become autonomous. Such unit families are clearly of immense importance among humans also. The bonds due to shared experiences and family affection may remain strong for many years, and over great distances. This general biological concept of a one-mother family is different from that of a 'nuclear family', as used by students of human societies, which refers to a group of two parents plus their one or more children, and including no further individuals whatever.

In a band of mountain gorillas, most individuals build a simple nest each night, though for the first three years or so the infant sleeps in the same nest as the mother. Among extant food-gathering societies of humans, a father, mother and children may sometimes go off together (for a period) as an independent one-mother family. But a wandering human food-gathering band is commonly a multifamily society, like that of the gorilla. Each family may build a temporary hut, to last for a few days or weeks. But after crop agriculture became important, when people had to stay in the same place to look after their fields and grain stores, permanent and more substantial huts became a normal feature. The wandering band changed into the society of a settled village. No doubt a hut or small house would sometimes be occupied by one unit family, who had a hearth and cooking fire of their own. As time went on, the mass and volume of the made house or home became increasingly large in relation to the mass and volume of the human members of the household; and the mass, number and diversity of other customary artifacts has kept increasing too. If a family has a car, that artifact alone will add another 500 to 1000 kilograms or more to the total of normal possessions.

In contemporary times, the persons who live together in the same house, apartment or hut commonly constitute one 'household'. And for many day-to-day purposes the group of individuals who are in fact members of one household – irrespective of relationships – is a more important social unit than one based solely on reproductive relationships. So it is felt that a human household requires recognition as a modified form of Level 7 society. The group of people who live together, and who breakfast from one kitchen, can be referred to as a 'one-kitchen household'. An ordinary Level 7 society is a one-mother family; but a different Level 7 society is the one-kitchen household. Many one-kitchen households are also one-mother family societies but some are not. And starting the other way round, only a minute fraction of one-mother families are also one-kitchen households, because of course the vast majority of one-mother families are in non-human species.

The society of a household often does consist of two parents and two or three children. But many are not like that: in a given small house or apartment there may be a couple without children, or two elderly sisters, or just one person; or one kitchen (being the place where one cooks or prepares food) may have to serve the needs of several or many unrelated people. This is notably so in institutional households, such as those of a hospital, a refuge for battered wives, a prison, an army unit, or a monastery.

Sometimes several unrelated persons, students for instance, share a small kitchen where they cook and wash up either separately or collectively. That group constitutes a one-kitchen household. Their other rooms may be close together, but in particular they are bound together, and have need to collaborate, through the sharing of a single kitchen. There may be a number of similar households nearby. In an army unit containing several hundred men there may be three kitchens: one for the officer's mess, another for the sergeant's mess and a large kitchen serving all the individuals of lower rank. In this situation there are three one-kitchen households, and they neatly correspond to differences in military status or rank.

Consider the case of two elderly sisters living together, in a small house, and eating from the one kitchen. They rank as a household. Perhaps at one time the same house had also contained the husband and children of one of the two sisters. During these social changes the house, the furnishings, the

Complexity and evolution

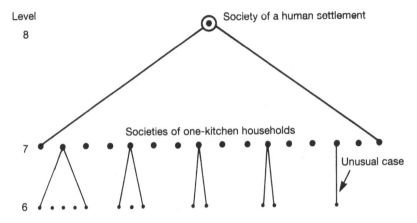

Level   Society of a human settlement
8

Societies of one-kitchen households
7

Unusual case

6

Individual persons, grouped according to households

Fig. 8.1. The societies of one-kitchen households, in relation to the levels above and below. Several private households of different size are analysed, including the unusual case where a Level 7 entity consists only of a single Level 6 entity.

cooking equipment and the garden around the house may have remained, apart from some changes, very much the same. But now the household consists of only the two sisters. If then one of them dies, and the other continues to live in the same house as usual, we are left with a 'household' consisting of only a single person. But the same house, furnishings, utensils, stove and garden are still there as before. In this situation it seems permissible to make a rare exception and to regard the single remaining sister as herself a one-kitchen household, as a modified member of Level 7, such as is normally composed of two or more members of Level 6 (Fig. 8.1). This usage can also be extended to other persons living alone, and eating from a separate kitchen. But it must be stipulated that the customary artifacts of a household must be present in addition to the one person. In the rare case of a person fleeing, alone, across country, sleeping under hedges and perhaps not breakfasting at all, he or she is regarded as a detached individual, who is not at the time a member of any household.

Older definitions of the word speak of a 'household' as consisting of those (plural) who live together in one house and compose a family. However, Carr-Saunders, Jones and Moser (1958) already recognised the possibility of the number of persons in a 'household' being only one. Indeed they reported that, in 1951, 11% of the private households did consist of only one person, while the commonest number of persons was two (as occurred in 28% of the households). Twenty years later, in 1971, the proportion of our one person households had risen to 18%. In 1987 in Britain as many as a quarter of the households contained only a single person; and in West Germany the proportion was nearly a third. Before the end of the century, over 30% of the households in Britain are expected to contain only a single person. But with each single-person household, there is at least a union of person plus household appurtenances, even if the household itself does not consist of person plus person.

In the previous paragraph, a house with its furnishings, etc. has been virtually allowed to fill the role of a person. This suggests an opportunity to mention, *merely as an aside*, an alternative general way of grouping things. One may think of the many hundreds of ants who inhabit a certain anthill; but might it be better to make a point of considering the ant society plus the made anthill as being the 'composite' entity of pre-eminent importance? Similarly one can consider the composite entity of a person plus their clothes and other possessions; a family plus its house, etc; and the society of a sovereign state plus all the personal and family artifacts and all the more general artifacts such as factories, railways and planes. That method of grouping (that kind of taxonomy) can give a more complete picture, but it is liable to involve more cumbersome and lengthy statements; and it would make it more difficult to do a clear-cut review of the mass of things, as later in the upper part of Fig. 9.2. We are not likely to forget that people commonly do wear artifacts (clothes) and live in artifacts (houses), and it is simpler and clearer to focus attention on the people themselves. However if one does start paying more attention to such 'composite' entities, there are some very interesting points to consider. Should a bacterial cell be thought of itself as a composite entity, with many of its molecules being in a sense 'artifacts' of the chromosome? If a family runs a farm where there are of course buildings, and where crop plants grow in somewhat altered (or made) soil, and where horses pull carts and ploughs, this farm is certainly an important and composite entity. And it is interesting to compare the nature of a leaf cell, which contains chloroplasts and mitochondria, with that composite farm. The chloroplasts have a role somewhat like that of crop plants, and the draught horses have a role somewhat like that of mitochondria. So is an ordinary cell already a 'composite' entity?

If a large old house is occupied by a number of families and others, living separately, then the number of one-kitchen households is to be reckoned from the number of kitchens. The same rule is applicable to a large hospital or an army unit. So the kitchen (the place for cooking and preparing meals) takes a place as one of the series of 'centres' or 'headquarters', along with mother, chromosome, and atomic nucleus. As a theoretical or accountancy arrangement – scoring the number of households by the number of kitchens – a kitchen can certainly be ranked as the 'centre' of a household. But especially in the many cases where one eats in the same place as the food is cooked, a kitchen often does have a real and conspicuous social role as the centre or hub of the whole household. Fig. 8.1 shows the position of such households in relation to the levels above and below.

Level 8: Human multifamily societies

The early species of humans, able to make stone tools, probably lived in wandering, food-gathering bands somewhat like those of the mountain gorilla. Later on it was also discovered how to kindle and use fire. Eventually our own species *Homo sapiens* emerged, and displaced earlier species. Multifamily food-gathering bands of *H. sapiens* eventually colonised every continent except Antarctica. Australia had been reached by 40 000 years ago, and possibly much earlier than that. 'Food-gathering bands' is a general term which can be used for those early human groups who had not yet got round to *producing* food by farming.

The first of three waves of humans to reach Alaska in North America, from Siberia, evidently arrived there before 12 000 years ago; and the advance continued so as to reach the southern tip of Chile by 11 000 years ago. The average speed of geographical advance through the Americas is thought to have been about 10 miles per year (Turner 1989). The linguistic result of the three separate entries from Siberia is well illustrated by Greenberg and Ruhlem (1992).

Perhaps it is of interest to mention that a similar geographical spread can be observed in the feeding habits of greenfinches. A family of greenfinches discovered that the kernels of the fruit of a small shrub called *Daphne mezereum* are an attractive food. During the first 100 years after the initial discovery in the Pennine region of the North of England, the new eating habit spread through Britain northwards and southwards at something like 2.5 miles per year (Pettersson 1956, 1959*a,b*, 1960*c*, 1961; Petterson & Pritchard 1988). That compares with the 10 miles per year progress of human foodgatherers through the Americas.

After crop agriculture became important, the society of the wandering band became transformed into the society of a settled village. More attention could be paid to the refinements of hut or house building. Normally there were domestic animals as well as farm crops. The amenities of village life developed, partly through baking and brewing, spinning and weaving, and pottery making. Some of these crafts were initiated at least twice independently, and perhaps several times, in different parts of the world.

One of the most important centres of origin of crop agriculture, in the Old World, was provided by the lands near the Tigris and Euphrates rivers and elsewhere in South-West Asia. And it may be mentioned that this general region is now also thought to be the centre of origin of the Indo-European family of languages as well as of some others. That family includes ancient Sanskrit and some modern Indian languages besides most of the languages of modern Europe. Colin Renfrew makes the interesting suggestion that while the early farmers slowly colonised Europe, seeking ever more land to cultivate, they took their diverging forms of Indo-European language with them. Indeed it seems the spread of farming caused the spread of the languages. And it appears that agriculture, with our first Indo-European language, arrived in Britain at a geographical speed of rather less than one mile per year (see, for example, Renfrew 1989; Gamkrelidze and Ivanov 1990). It seems quite understandable that the advance of the first crop-farmers across Europe should have been very much slower than the advance of those food-gathering bands down the Americas.

The early villages were multifamily domiciliary societies, and at first the society of each village was presumably autonomous. It was an independent Level 8 society. Nowadays of course virtually every settlement is fused politically with others.

We shall not here be considering persons grouped according to where they work, or by the kind of work they do, nor shall we be grouping people according to sparetime activities. We shall only be concerned with multifamily domiciliary societies, roughly the societies of settlements. But the enormous problem remains as to how much should be regarded as a single settlement. It is the same kind of problem we have met before at the levels of molecules and of cells. But we are now further up the natural hierarchy and can expect if anything an even more tantalising diversity of qualities and relationships than at the lower levels.

At Level 7 we made a special arrangement to recognise the one-kitchen household, in *Homo sapiens*, in addition to the one-mother families of the whole host of social species including *Homo sapiens*. And now at Level 8 it seems appropriate to make use of the well-known distinction between (1) geographical and (2) administrative or political criteria. If one drives through the English countryside on one of the ancient main roads, one may pass through a village, a village, a small town and another village. Having many fields between them, each may be a geographically distinct settlement.

The multifamily society of each is made up of the households of the central group of houses plus those of a few outlying farms etc. On a large-scale map such villages and small towns often show up as being clearly distinct. Each is occupied by its own Level 8 society. It may be fairly easy, from the map, to come to a practical conclusion as to how many such Level 8 societies there are in a certain area. The societies of these settlements have of course been distinguished *in a geographical manner*. The persons of the households of one small village are often known to each other fairly well. In a small town one knows fewer of the total number of people and in a city like Birmingham or London there is still less social cohesion. But from the *geographical* point of view, the whole of London seems to be occupied by one enormous Level 8 society. It appears that in other countries one can also recognise distinct Level 8 societies, distinguished in a geographical manner, in much the same way.

If on the other hand one is concerned with practical local government, the situation looks quite different. And the structural relations in different countries are very far from being uniform. *For administrative purposes*, London is divided up into a number of separate Districts, each administered by its own local authority, while in rural areas several towns and villages may be combined into one comparable administrative unit (Fig. 8.2).

It may be as well to point out that in Britain today, as elsewhere, there are a number of lower and higher tiers of local administration. There are many parishes, fewer local authority districts, and fewer still counties. Above counties, certain regions also have special administrations, and at a higher level still there are the formerly separate national units such as Scotland and Wales. Only the first three entities of that series seem appropriate to be taken as the possible base for the 'society of a settlement' in the administrative sense. One asks the question as to which of those first three tiers takes precedence as being associated with the greatest concentration of administrative functions. In Britain it is clearly the level of local authority district which takes precedence. Thus it is the population of each such districts which merits recognition as the society of a settlement in the *administrative* sense.

It is interesting to trace how the present system arose. Several centuries ago each parish commonly contained one village, or one main village plus a smaller hamlet. (Laslett (1968) suggested that in England around 1700 a parish contained on average about one and a half settlements.) The parson was the 'spiritual' head of the society of the village, and the local main landowner (later called the squire) provided a 'temporal' head. The society of the parish, in those days, was often a geographically distinct unit, in several

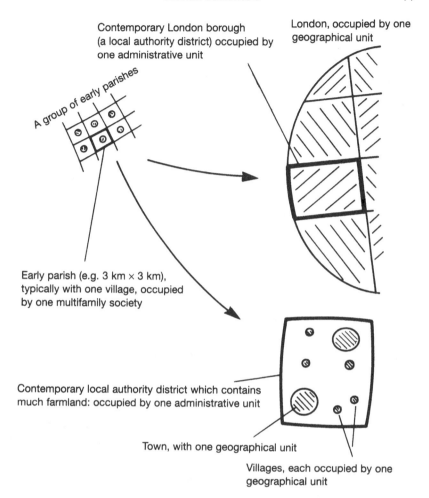

Fig. 8.2. Some of the historical developments affecting the societies of human settlements during recent centuries, as exemplified by changes in Britain.

respects it was an administrative unit, and the families of this multifamily society were often related. The number of parishes in England and Wales in the early nineteenth century, as cited by Thomson (1978), shows that the mean area of a parish was at that time just under 10 square kilometres, like the area of a square of which the side is roughly 3 kilometres or 2 miles. Such an area often would contain no more than a single village. We may note a couple of recently studied examples. Parker in *The Common Stream* (1975) has described the changing fortunes over many centuries of the people of Foxton in Cambridgeshire. The area of Foxton parish today is 7 square

kilometres. Blythe (1969) has collected biographies of many of the inhabitants of a village in Suffolk to which he refers as *Akenfield*, and the total area
of that was about 5.5 square kilometres. They both happen to be of less than
10 square kilometres, but no doubt many of the parishes containing barren
moorland have areas much above the average.

In the time of Queen Elizabeth I the society of a parish still appears often
to have been an effective Level 8 society in the administrative sense as well
as in the biological and geographical senses. But it is instructive to follow
the clue of the early development of social services with regard to coping
with destitution. The Poor Relief Act of 1601 gave each parish the legal
obligation to look after the destitute members of its society, using money
compulsorily collected at a standard 'rate' according to the value of property.
Despite various teething troubles the practice of such regular rate collections
proved to be socially useful, and it persisted for almost four centuries. But
already by 1723 the custom was officially sanctioned of several parishes
combining for the collection of rates and for putting the money to use, as
with a single 'work-house' for the destitute. For administrative convenience,
the process of amalgamation has continued, and many further local government functions have been added. In 1900 small towns, of only a few parishes,
often still collected and administered their own rates. But after recent reorgan-
isation the Local Authority Districts in Britain which collected and administered the general rates had – in the 1980s – populations which were often
around 100 000 and which ranged from about 10^4 to 10^6. And each of these
large groups of people, who share a common administration just as those in
an ancient parish once did, can be recognised as an administrative type of
Level 8 society.

The local authority of each district is headed by a mayor or leader, etc.
(elected) and by a town clerk or chief executive, etc. (appointed and paid).
Great urban areas are for convenience divided up into a number of separate
local authority districts. Thus the population of London, itself a geographical
unit, is divided into that of the thirty-two London boroughs, into thirty-two
administrative units. Fig. 8.2 shows this diagrammatically. One way a group
of small early parishes might develop was into the administrative unit of a
London borough. But elsewhere and more often a group of parishes
developed into an administrative unit composed of a number of geographical
units, each the society of a distinct village or town. Certainly the modern
local authority district does correspond administratively to the parish of
earlier times, and its population merits recognition as a form of Level 8
society.

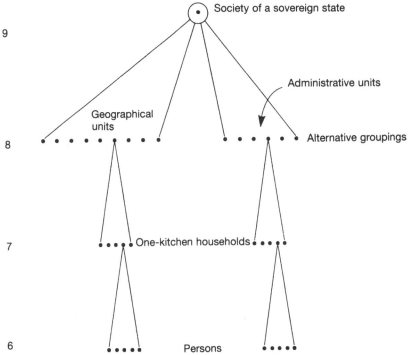

Fig. 8.3. Diagram showing the alternative ways of analysing the society of a sovereign state: either on geographical or on administrative criteria.

Sometimes a modern administratively defined unit contains a number of geographically defined units; and sometimes a geographical unit contains a number of administrative units (Fig. 8.2). Fig. 8.3 shows the alternative ways of analysing the society of a sovereign state.

It would be agreeable if other countries also normally had important administrative units with a population of around 100 000, and ranging from 10^4 to 10^6. But this seems not to be the case. Often the various regulatory functions of a British administrative unit are divided between somewhat larger and somewhat smaller units; and a special study would be needed in each case to decide which of the two units should take precedence. However Indiana and some other States in the USA do show approximately the same pattern as in Britain. In Indiana the various 'counties' (each with many important administrative functions) have an average population of about 60 000, and the range is from nearly a million down to 5 000. So in Indiana not only do the local government functions of a county approximately correspond to those of a local authority district in Britain, but also the size and range of the populations provide an approximate match. Clearly in Indiana the populations of each county can be ranked as an administrative unit corresponding to those in Britain. The average area of a county in Indiana is about 400 square miles.

Chapter 9

Human societies (second part)

This chapter will be mainly about the societies of sovereign states; about complex societies at Level 9. However, certain more general topics will be included.

In August 1993 the number of member states of the United Nations was 184 and as a first approximation the number of members of the UN can be taken as the current number of sovereign states in the world. Each sovereign state maintains order and governs within its own territory. It normally keeps a strict control of the movements of persons, other organisms, and artifacts (including money) across its borders. The populations of Britain, France, Germany, the United States and Japan are examples of Level 9 entities.

The society of each sovereign state is made up of the societies of settlements. And we must recognise that there is sometimes one or more important tier of government between that of our standard administrative unit and the society of the sovereign state. For instance each 'State' within the USA has its own influential government, but even California is not a *sovereign* state. It will be interesting to see whether, in the years ahead, something like a United States of Europe does emerge; or whether, on the other hand, the existing states co-operate more fully without actually fusing.

When, where, and how, did the early sovereign states arise? In the days of the first little farming villages, probably each community was more or less independent. An independent Level 8 society might be on good terms with some neighbouring villages and on bad terms with others. With a continuing and abundant supply of food, the size and number of early settlements increased. In a variety of different local ways, chiefdoms and city states developed, where the societies of a number of neighbouring settlements are seen as fused into a Level 9 society (into the society of a sovereign state). Daniel (1968) has coined the term 'synoecism' for the union of several towns and villages under one capital city, by deftly adapting the Ancient Greek

word from Thucydides, *synoecismus*. Perhaps its meaning might be broad-
ened slightly, so that the word synoecism can be used for the fusion of the
societies of settlements, to constitute the society of a sovereign state, even
when it occurred in earlier times and before any fully developed 'capital city'
had appeared.

Before the days of agriculture, those wandering food-gathering bands of
Homo sapiens which spoke the same language may have had 'a vague sense
of belonging'. The different bands of food-gatherers lacked formal organisation. But when the societies of contiguous agricultural settlements collabor-
ated closely, some form of central organisation tended to develop. Sometimes, as in the Tigris and Euphrates regions in South-West Asia, the main
village of the group became a 'town' or 'city'. Archaeologists speak of the
independent 'city states' centred on the cities of for instance Ur or Lagash.
(At first they were independent, though later they were fused into various
larger states and empires.) In each case there was a compact settlement of
some thousand or more inhabitants (the 'city'), around which would be a
number of villages and hamlets occupied by those who tended the fields and
flocks of the territory of the state, and whose place of work was too far from
central settlement for them to be able to return to sleep there each night. A
chief, war leader, priest or priestess, assisted by others, came to exert a domi-
nating influence on the general activities of a group of linked settlements.
Insofar as such persons, who controlled events, spent no time on production
of food or pottery etc., they were functioning as full-time administrators.
Such differentiation is an essential feature of Level 9 entities. One of the
ways of deciding, archaeologically, whether an important chief or priest was
in fact operating in an ancient settlement, is to discover whether there is
evidence of a distinctive and often larger building to act as his headquarters.

It appears that the emergence of Level 9 entities, sometimes occurred
already in neolithic times, that is before the bronze age. Some examples may
be mentioned. Çatal Hüyük in Anatolia, Turkey, has been excavated by James
Mellaart, of the University of London. Scarre (1988) mentions that Çatal
Hüyük was founded about 9000 years ago, and became the largest neolithic
site in the Near East. Its closely packed houses occupied 13 hectares. It is
famous for its impressive mud-brick architecture, craftsmanship and art. Bar-
bara Bender (1975) notes how Çatal Hüyük developed as an important trading
centre for materials such as obsidian (which was mined not far away) as well
as for greenstones, lava, limestones, alabaster, marble, flint, copper and large
cowries (from the Red Sea). At or before 8000 years ago one finds evidence
of priestly quarters, and of a great bull cult. This occupational differentiation
with priests, miners and traders, in addition to farmers, is characteristic of

Level 9 societies. It suggests strongly that a number of smaller settlements were linked to the larger central one, and that a 'society of a sovereign state' had indeed emerged at Çatal Hüyük by some 8000 years ago. That is during the neolithic period, and long before the advent of the bronze age in the region concerned. Furthermore Colin Renfrew (1989) suggests that the region around Çatal Hüyük may have been the actual centre of origin of the Indo-European family of languages. It seems not unlikely that by 8000 years ago there were also some other Level 9 societies in the Near East.

Coming nearer home, to Britain, there seems to be evidence of multisettlement Level 9 societies here too during the neolithic stage of cultural developments and before 4000 years ago. In Wiltshire there is an enormous man-made mound known as Silbury Hill. There are also the carefully arranged stones of Stonehenge. Colin Renfrew is among those who have argued that in each case the amount of heavy constructional work involved points to the centrally directed co-operation of the inhabitants of quite a number of different settlements. Hence the group evidently functioned as a Level 9 society. Euan MacKie (1977: 155) shows the plans and the reconstruction of two of the large wooden houses, of neolithic age, such as those from which the organisers of Woodhenge or Stonehenge may have operated. The larger was as much as 40 metres across, being a circular building, perhaps with a small open courtyard in the centre. So it was several times the diameter of a prosperous iron-age farm-house, such as was used in Britain nearly 2000 years later. It was a large building suitable for the headquarters of an early sovereign state. While in South-West Asia such states may date from about 8000 years ago, the same level of organisation was represented in Britain and doubtless many other regions by some 4000 years ago.

Though it is not always realised, in most regions there were certainly Level 9 societies for several or many centuries before the advent of reading and writing in the region concerned. For instance prior to about 2000 years ago when the Romans impinged, there were at least 16 non-literate tribes in England and Wales; and the territories of some if not all of these tribes can be regarded as the territory of a single Level 9 society. Barry Cunliffe has constructed the map reproduced as Fig. 9.1 (Cunliffe 1974).

By about 5000 years ago in the Tigris and Euphrates valleys of South-West Asia and in the Nile valley of Egypt, and later elsewhere, Level 9 societies had developed ways of recording both words and numbers with writing and arithmetic. Both seem to have developed as aids to administration and store-keeping, (Childe 1942). Forms of writing, in Level 9 societies, developed independently in several different regions of the world. Once writing is present, one can speak of a literate civilisation. With the beginning of writing,

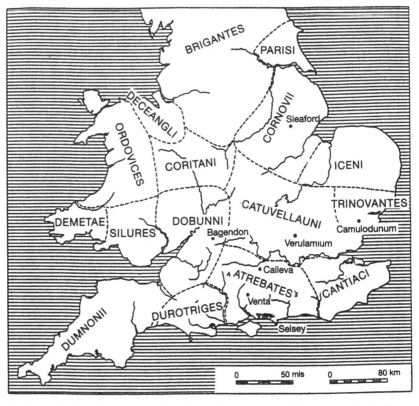

Fig. 9.1. Map showing some of the tribes in Britain when the Romans arrived. Many have heard of the Queen of the Iceni, Queen Boudicca, so the Iceni, for instance, evidently did constitute a Level 9 society. (From Cunliffe 1974.)

in each area, 'history' in the strict sense begins and soon more can be learnt about life in the ancient Level 9 societies from written records. In contrast to the way of life in many of the independent human Level 8 societies of earlier times, the features of early literate civilisations include:

(1) Full-time occupational differentiation among persons even of the same age and sex

(2) Considerable economic differentiation into richer, medium and poorer

(3) Considerable social differentiation, with those most respected and privileged at one end of the scale, and often a condition of unfreedom (e.g. slavery) among the less privileged at the other end.

While slavery is nowadays less conspicuous, we may note that these three forms of differentiation – occupational, economic and social – have, to a greater or lesser degree, been conspicuous features of all known Level 9 societies, ever since.

One or two thousand years ago, there must have been a very large number of sovereign states in the world. Even 300 years ago in Europe, there are said to have been some 300 virtually independent states. Today however the total number in the world is about 184.

Turning to the evolution of early multicellular motile animals, one does not know how many cells were present in the first organisms which became sufficiently differentiated as to possess nerve cells. They may have had more or less than a thousand cells. At a higher integrative level it will be interesting to try and work out how many people were usually present, in a group of politically linked settlements, before sufficient internal differentiation occurred for full-time administrators to arise and thus for the group to graduate into the society of a sovereign state. Some time ago it was suggested that the necessary threshold was around $10^{3.5}$ or 3000 (Pettersson 1960*a*). Later the estimate may need to be modified, perhaps by less than a factor of three. For the time being, pending further information, we may think of the early Level 9 societies – each being an independent and integrated group of Level 8 societies of settlements and with a moderate degree of internal differentiation – as containing around $10^{3.5}$ or 3000 people. Such a sovereign state, centred on Çatal Hüyük had apparently evolved on the Anatolian plateau by about 8000 years ago, though later it may have contained 10 000 or more people.

Although they were very different in size and power, such early Level 9 societies must be reckoned as of the same general category as the societies of sovereign states today. Each is a politically independent entity and each is composed of the societies of a number of settlements. One of the smallest today is that of Iceland which in 1974 comprised 216 000 persons, far more than in those early states but less than the number in some London boroughs. The largest Level 9 society is that of the People's Republic of China, with just over a thousand million people. Taking the number in an early small state as 3000, that means that China is greater by a factor of at least 300 000 or $10^{5.5}$. This is a large factor, though less than a million. The question arises: is a factor of about a million too great to have among entities which are ranked as of the same major integrative level? The ratio of the numbers of people will of course be similar to the ratio of the mass, or biomass, of the states concerned. Referring back to Fig. 6.2 we see that at many levels the ratio, between the heaviest and the lightest, is far more than a million. On that diagram only atoms have a ratio smaller than that of societies of sovereign states. From the point of view of the theory of major integrative levels, therefore, the societies of small early states and of large modern states differ *less* among themselves, in numbers and mass, than the members of many other major levels.

How many independent societies?

Quite apart from the number of human individuals on Earth, or the number of Level 9 societies, it is interesting to focus attention for a moment on the number of independent human societies there were, at various times, and on how many there are likely to be in the future.

At one time there were no humans. When a multifamily band of some early species first started to make stone tools, we may say that from then on there was at least one independent human society. As the newly acquired habit of stone tool making spread, generation after generation, the number of independent human societies may have increased to two, to four, to eight and so on. Later there was a long and fairly slow rise in numbers of independent human societies, as evolution proceeded from species to species and as different regions were colonised.

It is convenient to review the situation at about the time of the inception of food production by farming, or say at 10 000 years ago. (Farming commenced before that, but by 10 000 years ago it was still in an early phase and was still confined to a very small part of the globe.) It has been estimated that the world population, i.e. the maximum number of pre-agricultural humans, at about 10 000 years ago was 5 million (Ehrlich & Ehrlich, 1970). If at 10 000 years ago the average size of the independent human societies (mainly Level 8 but some Level 7) was 25 individuals, that would imply

$$\frac{5\,000\,000}{25} = 200\,000 \text{ independent societies}$$

So before the beginning of agriculture our ancestors lived in many thousands of independent societies but the social evolution triggered off by the inception of agriculture has produced Level 9 societies. There are now virtually no independent Level 7 or Level 8 societies. Despite the much greater number of people, there are today only something like 180 independent societies being roughly the number of members of the United Nations. So the number of independent societies has declined to barely a *thousandth* of the number there was before agriculture.

The number naturally fluctuates. We may wonder whether the long-term tendency towards fewer independent societies will continue. Ten millennia hence, could there be expected to be say 2000 independent societies, or only 10 or 2 or 1 ... or none?

Complete hierarchy using gravimetric triangles

Now that we have taken the trouble to comprehend the idea of households being a special form of Level 7 society and to recognise both geographically and administratively defined settlements, it is possible to plot an example or model of the complete natural hierarchy as in Fig. 9.2. The idea of special gravimetric triangles was introduced in Chapter 3. The portion of the hierarchy which runs from a person (a child) down to an electron, is repeated from Fig. 6.3. Now the child can be seen among the other members of a family household. In fact the same family model has been used here as in Fig. 3.4. At the level above, the household societies make up a Level 8 society (a geographically defined one is shown); and higher still these societies of settlements make up the society of a sovereign state. The approximate mass, or biomass, of the one here plotted (some $10^{12.5}$ grams) is roughly that of the United Kingdom including dependencies. While 'society of a sovereign state' is a precise term, another useful but less precise synonym is of course nation.

A few years ago the different nations of the world ranged in biomass from about 10^{10} grams for Iceland to about $10^{13.6}$ grams for China. Early in 1976 the world population passed the 4000 million mark. Taking the approximate average mass of individuals as 50 kilograms (to make use of a convenient round number), that suggests that the total biomass for *Homo sapiens* was then about

$$4000 \text{ million} \times 50 \times 10^3 \text{ grams}$$
$$= 200 \times 10^{12} \text{ grams, or } 10^{14.3} \text{ grams}$$
$$= 200 \text{ million metric tons.}$$

This also is plotted in Fig. 9.2. And if the true average mass were known, for that date, the total mass would plot at virtually the same position as shown on the present diagram. (With many individuals being children, that 50 kg is clearly too high an estimate.)

When considering Fig. 9.2 it should be realised that a slightly different diagram of this kind could be constructed for many of the different atoms in each of the different people on Earth. However, the diagrams would all have a strong family resemblance, sufficient for any of them, including Fig. 9.2, to be taken as a useful example or model. The diagram enables the quantitative picture of certain relationships to be obtained between natural

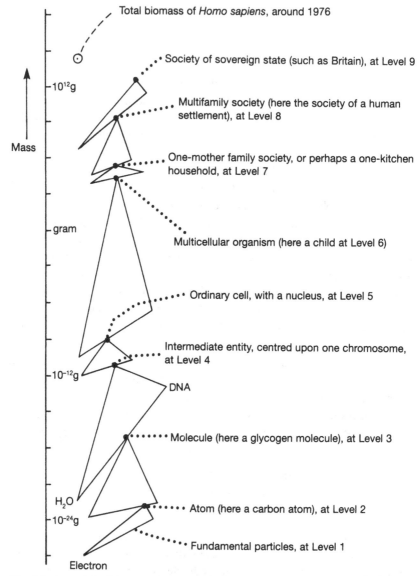

Fig. 9.2. A model of the complete natural hierarchy. It is an extension of Fig. 6.3, and now also shows the three social levels.

entities as far apart as fundamental particles and nations. The diagram brings together data which have seldom if ever been considered together before.

Fig. 9.2 provides an unusual insight into gravimetric relationships. When considering it one must remain aware of the vast host of qualities and functional relations, of the various entities, which are not shown on the diagram – just as a chemist for instance remains aware of the host of other properties of atoms and molecules, when he discusses atomic and molecular weights.

Having newly classified entities into the levels numbered from 1 to 9, these categories can now be used to make studies which also involve aspects of time, number and mass. Thus Chapters 10 to 14 lead to several completely new quantitative conclusions of wide validity. And these are then summarised in Chapter 15.

Chapter 10
Acceleration in evolution

Evolution

It was in 1859 that Charles Darwin published *The Origin of Species*. He proposed that all existing species of animals and plants had evolved from earlier and different species; and that 'natural selection' had helped to produce the changes. That set the topic of evolution squarely upon the agenda. Fifty years later probably most professional biologists had come to agree with the idea, while today it is generally accepted.

Let us briefly review the ancestry of ourselves and the other mammals, aided by Fig. 2.4. Perhaps somewhere around 2000 million years ago, our ancestors became cells with nuclei; around 1000 million years ago we became multicellular organisms with internal differentiation. Then about 500 million years ago we became early vertebrates, as fish. The 'backbone' was however at first made of cartilage, only later of true bone. Then followed a stage as amphibians, where the adults have four legs and have lungs for the breathing of air when on land. Frogs are our best-known amphibians today. Later the tadpole stage was cut out and we became reptiles, lizards. Then some 200 million years ago some creatures started to bear their young alive and to suckle them with milk – as early mammals. This also, of course, saw the beginning of social family life, at Level 7. Careful geological and palaeontological studies have helped to establish this sequence. It has been aided by isotopic dating, as discussed in Chapter 4.

Acceleration

Fig. 2.4 gives approximate dates for the initial emergence of entities at each higher integrative level. The same information can be set out as in Fig. 10.1.

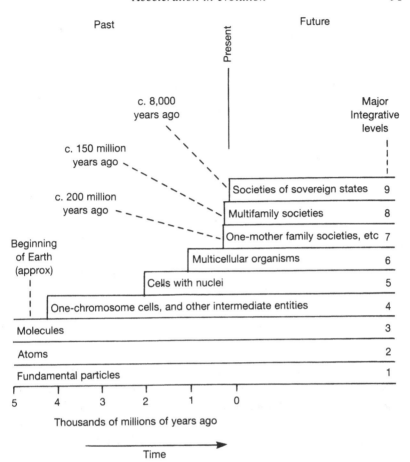

Fig. 10.1. There is a horizontal axis of time. The different major integrative levels are shown as successive steps up a staircase. The staircase becomes steeper as one goes up, indicating that there has in fact been acceleration in evolution. (For convenience of draughtsmanship, it has been assumed that organisms and societies will persist for some time to come.)

There is a horizontal axis of time, and the successive 'levels' are shown as a rising flight of steps. In so far as the staircase becomes steeper, going upwards, this is evidence of *acceleration* in the evolution from level to level. The same series of estimated dates is set out in Table 10.1, and later on margins of error will also be mentioned. In the right hand column of Table 10.1, it is seen that over the biological range, from Level 4 to Level 6, evolution took some 3500 million years; whereas over the social range, from Level 7 to Level 9, evolution had accelerated to take a mere 200 million years.

Table 10.1. *Estimated date and spacing between dates of the first appearance of each of the major integrative levels of the biological and social ranges*

Major integrative level	Kind of integrated natural entity	Estimated time (years ago) of initial appearance	Difference (a)	Difference (b)
9	Society of sovereign state	8000		
			150 million	
8	Multifamily society	150×10^6		200 million
			50 million	
7	One-mother family	200×10^6		
			800 million	
6	Multicellular organism	1×10^9		
			1 billion	
5	Cell with nucleus	2×10^9		3250 million
			2.25 billion	
4	Intermediate entity	4.25×10^9		

Table 10.1 shows up the one glaring exception to the generalisation that each successive step upwards is steeper for it suggests that it took three times as long to reach Level 9 from Level 8, as it took to reach Level 8 from Level 7. It will be realised that this is only a minor discrepancy when one observes the whole sweep of evolution shown in Fig. 10.1 which demonstrates acceleration in evolution very convincingly.

Another way of studying acceleration is through Fig. 10.2 and Table 10.2. In the first pair and in the second pair of 750-million-year periods there was only one step up, while in the third pair there were four.

Gravimetric acceleration

Stepping-up, so to speak, to a new and higher integrative level – for instance, when the first multicellular organism evolved from a one-celled ancestor – certainly represents a major innovation, thus the first new members of each successive level will be referred to as 'innovatory entities'.

The approximate times of appearance, on Earth, were listed in Table 10.1. These medium estimates are repeated in Table 10.3 but on each side there is now a suggested margin of error. The same table includes supposed values for the mass of each kind of innovatory entity, together with similar margins

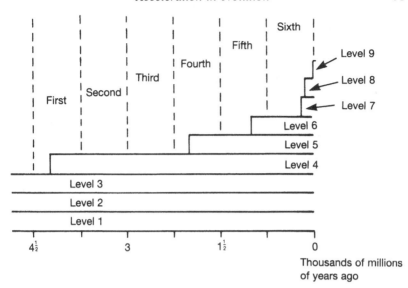

Fig. 10.2. A version of the evolutionary staircase diagram. Successive periods of time, each of 750 million years, have now been demarcated to help illustrate the acceleration. It can be seen that no steps up occurred during the second and third periods, but that there were as many as three steps up in the sixth period.

Table 10.2. *The frequency of step-like ascent, to membership of a higher major integrative level, during the six equal periods of time*

Periods of 750 million years	New entities emerging during the period	Number of kinds of new entities
Sixth	Three kinds of societies	3
Fifth	Multicellular organisms	1
Fourth	Cells with nuclei	1
Third	None	0
Second	None	0
First	Intermediate entities	1

of uncertainty. Soon we will plot these dates, but something should perhaps first be said about the magnitude of the various mass estimates.

Mass of innovatory entities

Possibly the critical population size for the emergence of a Level 9 society was about $10^{3.5}$ or roughly 3000 (Chapter 9), though the early state centred on Çatal Hüyük may at its zenith have later had 10 000 or more members.

Table 10.3. *The acknowledged margins of uncertainty about the mass, and the time of appearance, of innovatory entities at Level 4 to Level 9*

Major integrative level	Approximately estimated range of mass of innovatory entities, within which the true mass is likely to lie	Approximate factor between the central values of mass[a]	Approximate range of date of innovatory entity, within which the true date is likely to lie (years ago)	Very approximate doubling time, in years, with regard to mass of innovatory entities
9	450 – 150 – 50 tonne		10 – 8 – 6 thousand	
		1.5×10^4		
8	100 – 10 – 1 kilogram		200 – 150 – 100 million	12 million
		10		
7	10 – 1 – 0.1 kilogram		250 – 200 – 150 million	
		10^8		30 million
6	100 – 10 – 1 microgram		1.3 – 1.0 – 0.7 billion	
		10^4		
5	30 – 1 – 0.03 nanogram		2.5 – 2.0 – 1.5 billion	75 million
		10^9		
4	10 – 1 – 0.1 attogram		4.5 – 4.25 – 4.0 billion	75 million
		Geometric mean = about 10^5		

Using the convention of an average personal weight of 50 kilograms (a rather generous estimate), that suggests something like 150 million grams, or 150 tonnes for the mass of the first society of sovereign states. But as we shall see, a several-fold error is this estimate would not affect the general conclusion. Among the earliest mammals, the first one-mother family may have been about 1 kilogram or less, and the first multifamily society perhaps 10 kilograms or less.

Reviewing the size and hence the weight of various small Level 6 entities – that is of multicellular organisms with at least moderately good internal differentiation – 10 micrograms seems a likely figure for the first such entity. Even some adult insects such as the male louse *Enderleinellus*, may weigh as little as 5 micrograms. Other animals of between 1 and 10 micrograms include many free-living rotifers. And some plant and fungal multicellular

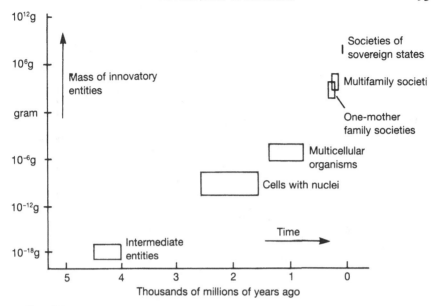

Fig. 10.3. These 'uncertainty rectangles' use the information in Table 10.3. It is considered very likely that the true position of any innovatory entity lies somewhere within the appropriate rectangle, even if not at its centre.

organisms weigh less than 1 microgram. So an estimate for the first multicellular organism at 10 micrograms seems by no means too low. Perhaps the first cell with a proper nucleus weighed about 10^{-9} gram or one nanogram; though many much lighter cells with nuclei are known today. And judging from the size and weight of very small modern viruses, the first intermediate entity may have been around 10^{-18} gram (Douglas 1975).

Plotting masses of innovatory entities against time

Fig. 10.3 plots 'uncertainty rectangles', showing the supposed margins of error, and using the data in Table 10.3. This already shows a distinctive trend, with regard to the arrangement of the rectangular blocks. Then Fig. 10.4 uses the medium values for both mass and time. We see at first an ascending straight line, suggesting simple exponential acceleration with regard to mass, while later the acceleration is faster still.

One of the ways of describing acceleration is to mention the doubling time. If a car is accelerating, and for an instant is travelling at 30 miles an hour, a passenger with a stopwatch could note the time taken for the speed to double to 60 miles an hour. And the time taken for the speed to double can be called

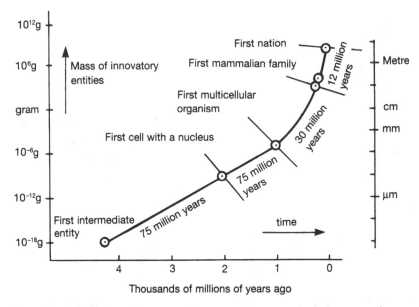

Fig. 10.4. The doubling times, with regard to innovatory mass during successive periods of biological and social evolution; simple exponential acceleration is indicated at first, but later the acceleration becomes faster still. From mass one can deduce approximate volume. Thus the scale on the right gives the length of the side of a cube of water, having about the volume and mass concerned.

the doubling time. Unlike a journey by car, biological and social evolution extends over some 4.25 thousand million years. The progression of mass values is best displayed with logarithmic plotting. But it has been found that to calculate the approximate doubling times happens to be one of the best ways of describing this acceleratory progression of the values of mass of innovatory entities. Fig. 10.4 displays the findings. Thus for the early part of the curve, up to the time when multicellular organisms appeared, the doubling time was apparently about 75 million years. But the doubling times then became shorter. Between the first family society and the first society of sovereign state, the doubling time was only some 12 million years. In due course these estimates will be revised a little, but the enormous differences between them will remain.

An aside on mass, volume and length

Since it is easier, for most of us, to visualise lengths and volumes rather than weights, it occurred to me to put a logarithmic scale of lengths on the right side of Fig. 10.4. Each such length represents the side of a cube of water having approximately the weight indicated on the left-hand scale of mass directly opposite (thus gram is on the same level as centimetre) and most living systems have a density which is not very different from that of water. It can be seen that the first cell with a nucleus is suggested to have weighed one nanogram, and to have had a volume like that of a cube of water with a side of about 10 micrometres. The first mammalian family, at 1 kilogram, would have had the volume of a cube of water with a side of about 10 centimetres. Even the society of the first sovereign state could have had a volume totalling that of a cube of water and with a side of about 5 metres.

Considering the immense differences, in structure and functioning, between an early nation and an early intermediate entity, it is perhaps surprising that the factor between them, on this linear scale, is not greater than here indicated.

Combining the two lines of evidence

Early in the chapter we found evidence of acceleration in evolution, in that the staircase showing the ascent from one major integrative level to the next level above became (in general) progressively steeper. Now in Fig. 10.4 we find evidence of another kind of acceleration, that of mass of innovatory entity. At first there was simple exponential acceleration, but later the acceleration became faster still. So long-term evolution has been subject to pronounced acceleration, and accelerations of two different forms.

So far as is known, this is an entirely new contribution to scientific and historical thought. J. M. Roberts (1976) is among those who notice that the acceleration of cultural change commenced before the beginning of agriculture; Leakey and Lewin (1978) notice the same point. Wilson (1975) includes data which show an acceleration in increase of brain size among our ancestors, during a portion of the last 10 million years. To all of us, hitherto, this acceleration of change has seemed a mainly human phenomenon, confined to our own lineage. Now, however, we find clear evidence of a general long-term acceleration throughout virtually the whole period of biological and social evolution. This new knowledge of acceleration in evolution, before human times, is an unexpected discovery.

Chapter 11
Further allied accelerations

As our ancestors and others ascended the natural hierarchy, from Level 4 to Level 9, there was acceleration, as established in Chapter 10. It will now be interesting to try and relate this newly discovered *evolutionary* acceleration to two other forms of acceleration about which a number of studies have already been published.

One refers both to acceleration in the increase of the number of humans on Earth and to acceleration in human cultural innovation. It will be found that a coherent picture does emerge, leading to what may be called 'the law of ever faster change'. The topics of demographic increase and cultural innovation stand somewhat apart from our central study of the natural hierarchy. But it does seem worthwhile to notice these processes, if only briefly.

The difficulties resulting from the increasing numbers of people are forcefully discussed by, for instance, Parsons, (1971) and Ehrlich and colleagues (e.g. 1970, 1978), while another American writer, Toffler (1971), provides a moving and well-documented account of the increasing rapidity of social change.

Method of plotting

Figures 11.1 – 11.3 all use the same special axes, which have been found convenient for the data under consideration. The horizontal axis is a logarithmic scale of chronological time, referring to the number of years before AD 2000. Thus the five markers on the right of the horizontal axis refer to the dates 1999, 1990, 1900, AD 1000 and to 8000 BC. This arrangement gives plenty of space for the numerous details known about the very recent past, and at the same time it permits entries for the early evolutionary developments of several thousand million years ago. The vertical axis shows rate of acceleration, as indicated by shortness of doubling time. It should be noted that this axis of doubling times (as drawn) ranges from one year to 10^9 years,

98

while the horizontal axis has a range from one year to rather more than 10^9 years.

The information about biological and social evolution, shown in Table 10.3 and Fig. 10.4, has been transferred to Fig. 11.1, where each of the entries to do with evolution has been labelled with an E. There are several different ways in which this transference of information could have been made. By considering the curve of Fig. 10.4 as made up of a large number of very small portions – and then transferring each of these separately – one could obtain a version which approximates to a continuous smooth curve. Or one could transfer just the four large portions for which the doubling times are marked in Fig. 10.4. With the time axis used in Fig. 11.1, however, the last portion (with a doubling time of 12 million years) would get pulled out to an abnormally great length, so it is convenient to break this last portion up into four, and in such a way as to obtain entries of roughly the same length. Thus in the new diagram there are now six horizontal entries to do with evolution. One can visualise how a curved line could replace these six entries, if a curved line were required, but for the present purpose it is quite adequate just to use a series of not-too-long horizontal entries. The last main portion of the curve of Fig. 10.4, with an overall doubling time of about 12 million years, has been broken up into portions which themselves evince doubling times of about 15 million, 3 million, 1 million and 150 000 years.

Demographic acceleration

The entries about numbers of people, in Fig. 11.1 are labelled D for demography. Starting from 2 million years ago, approximate estimates were obtained of the total number of human individuals who were alive at different times. Latterly of course only *Homo sapiens* was present. Doubling times for suitable periods were calculated and the results entered in Fig. 11.1. Around a million years ago, the doubling time itself was probably around 1 million years. As with mass of innovatory entities (Chapter 10), the increase in number of humans has, at least recently, been faster than a simple exponential increase. From a figure of 3000 million in 1960, the world population will probably double by the year 2000, which means a doubling time of as little as 40 years. The systematic shortening of doubling times from about a million years, to a mere 40 years, is an obvious case of change which is faster than simple exponential acceleration.

We may state the conclusion formally. The number of human individuals has been increasing in an acceleratory manner, and during recent millennia the acceleration has been faster than simple exponential acceleration. Over

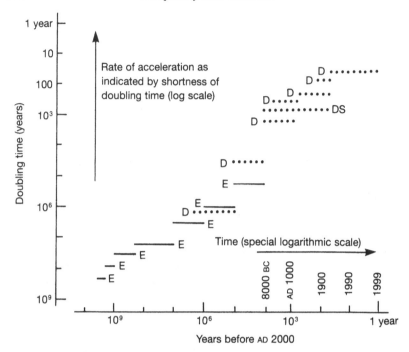

Fig. 11.1. The progress of evolutionary and demographic accelerations, using a special time scale. Given time and patience, the smooth curve of Fig. 10.4. could have been transferred to here, but instead it has been transferred in sections. DS refers to the populations of maximum settlements (please see text). E, evolutionary; D, demographic.

the past million years the doubling time has shortened from around a million years to about 40 years.

The entry in Fig. 11.1, labelled DS, refers to the populations of maximum settlements (Pettersson 1960*b*). Before the beginning of agriculture, relatively few people lived together, but during the past 10 000 years, as revealed by archaeology, the size of settlements has been increasing. If one puts together estimates, by different authorities, of the number of people inhabiting whichever was the world's largest settlement, at this or that period, a generalisation emerges which can be plotted as DS and which can be stated formally as follows. Since 10 000 years ago the population of the world's largest settlement (village, town and then city) has increased in an acceleratory and approximately exponential manner, with a doubling time (until recently) of about 600 years.

Evolutionary and demographic acceleration

Hitherto we have refrained from commenting on the general appearance of Fig. 11.1 as a whole. The E entries refer to increase in mass of forefront innovatory entities during the ascent from major integrative Level 4 to major integrative Level 9. *Homo* may be regarded as the forefront zoological genus now alive, and the D entries refer to increase in the number (*or* indeed the total mass) of individuals of this forefront genus. There is some overlap between the ranges of the evolutionary and the demographic entries in Fig. 11.1. But the strikingly obvious and quite unexpected situation is that the E and D entries, as plotted, seem to fit together as members of a single linear band. Going along the series, doubling times have shortened from a maximum value of 75 million years, down to a minimum value of 40 years. This unexpected link-up can be recorded as follows. As chronological time advanced, the shortening series of evolutionary doubling times is seen to be continued (up to the present), as a broadly similar series, by the demographic doubling times.

Cultural aspects

The process of innovation or invention of a new cultural element is obviously different from the subsequent process of the spread of that cultural element, by copying, among the population concerned. Such innovations, followed by the spread of the new cultural element, can of course be recognised in some other species of mammals as well as among humans, and in birds as well as in mammals. The rates of replicatory cultural advance or cultural diffusion are found to be relatively uniform. Indeed the doubling times, whether more or less than 10 years, are relatively unaffected by the level, in the social range, of the species concerned, while even among humans the rates are relatively unassociated (until recently) with chronological time.

However, the situation is quite different with regard to the rate of innovation or invention of *new* cultural elements. Besides our genetic heritage, humans today possess an immense cultural heritage. Developing a body of knowledge and practices which are passed on from individual to individual, usually long after birth, has been a forefront development of humans and human societies. There is a general feeling in many countries today that cultural change is accelerating and there have been several studies of the rates of cultural innovation.

Palmer (1957) carried over the quantitative methods from the exact sciences to the study of archaeology and published careful estimates, for various periods, of:

(1) The number of different materials used by man
(2) The number of occupations involving special arts and techniques which were pursued
(3) The maximum speed of transport by mechanical means
(4) The complexity of man-made objects and the degree of skill and knowledge required to produce them.

The four series, when plotted, show similar curves with similar acceleration; and the four series may be combined to give a single index of general cultural advance. Between 30 000 years ago and 10 000 years ago this index has a doubling time of about 100 000 years; while between AD 1750 and 1950 the doubling time has shortened to about 65 years. The more important conclusions about rates of cultural innovation, from the work of Palmer and others, are plotted in Fig. 11.2.

Somewhat comparably to the work of Palmer, Sir Ieuan Maddock has plotted the progress of recent technological innovations up to about 1975. He deals with:

(1) Speed of travel, noting that there are now spacecraft
(2) Communication, noting TV satellites
(3) Killing, noting intercontinental ballistic missiles
(4) Data processing, noting computers and microprocessors.

Combining the four similar and acceleratory progressions, one may obtain a single index of recent technological facility. Between 1800 and 1900 this has a doubling time of about 150 years. But between 1950 and 1975 the doubling time has shortened to less than 2 years, perhaps about 1.3 years.

It was the pioneer work of D. de Solla Price which drew attention to the acceleratory, indeed exponential, growth of scientific information (1956). The number of scientific journals founded, throughout the world, between 1750 and 1900, had a doubling time of about 15 years. The numbers of scientific and technical papers published each year in various disciplines, during the present century, have similar or shortened doubling times, while latterly the overall doubling time is estimated at about 9 years.

One of the most distinctive and influential developments in the forefront of cultural advance, of recent years, has been the construction of integrated electrical circuits, on small chips. It was pioneered in the United States. During the first eighteen years, following 1959, the maximum number of components per integrated circuit has in fact had a doubling time of only

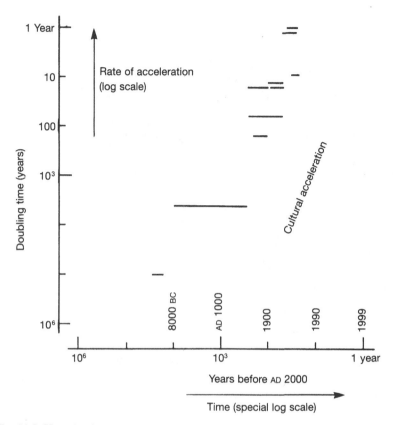

Fig. 11.2. Here the doubling times of important cultural innovations – using the data of Price, Palmer and Noyce – are plotted along the same special time scale.

one single year (Noyce 1977), and that provides the highest entry in Fig. 11.2.

Reverting to the several different forms of activity studied by Palmer, to the several different topics studied by Maddock, and noting as well the studies of further single topics, we may combine these results by stating that particular sequences of innovation, in a given field, usually have an acceleratory phase.

The display of data in Fig. 11.2 may be further described as follows. Over a long period the plotted doubling times for a wide variety of different kinds of important innovatory cultural advances, among humans, form a coherent and fairly narrow band. And over the past 30 000 years the rates of acceleration, in a variety of forefront innovatory cultural developments, among humans, have fairly consistently increased. Cultural changes have been

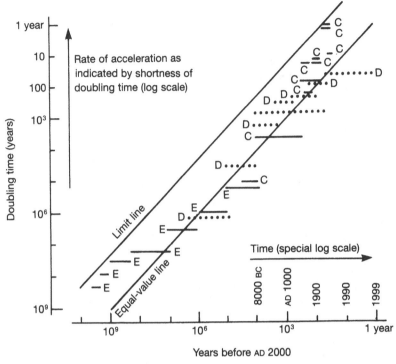

Fig. 11.3. Here the data of Figs. 11.1 and 11.2 are plotted together. C, cultural; D, demographic; E, evolutionary.

getting faster. The doubling times have shortened, for instance, from 100 000 years to 1 year.

Demographic and cultural advance

Earlier we noted that the trend of demographic accelerations does approximately continue the trend of evolutionary accelerations. Let us now compare the demographic and cultural data. They are plotted together in Fig. 11.3. some of the more recent cultural doubling times have certainly been much shorter than the recent demographic doubling times, but that is only in the last few centuries. Over almost all of the past 30 000 years the trend of quickening increase in invention has been similar to the trend of quickening increase in population. With regard to humans there has been a broad similarity, over a long period (until recently), between the doubling times of demographic increase and the doubling times of innovatory cultural advance.

Any termination in prospect?

Before a general discussion of Fig. 11.3 let us pause for a moment and con-sider future prospects. Most of this chapter shows how several processes have been getting ever faster, more or less in unison. Will it go on for ever? Probably not. If, for instance, the world population one day stabilises, there would then be no demographic acceleration, and the demographic doubling-time entry would drop to below the bottom of the page. But cultural inno-vation could well go soaring on. Thus we may predict, with some confidence, a definite *termination* of the close association between the demographic and cultural trends. Even merely by looking at Fig. 11.3, one can see the C entries drifting rather fast upwards, and already becoming somewhat divorced from the trend of D entries.

General review of accelerations

The data we have obtained on evolutionary, demographic and innovatory cultural accelerations are all plotted together in Fig. 11.3, with the three cat-egories being labelled E, D and C. Taken together the entries form an elon-gated band some five or six times as long as it is broad. We have made the unexpected discovery that there has been a general coherence of the forefront doubling times in biological, social and cultural evolution and in demographic increase. The axis of the trend points upwards, leading to what might be called 'the law of ever faster change'. The doubling times of these processes have systematically shortened – in fact from 75 million years down to a single year.

Looked at numerically, the acceleratory changes of today are seen as a continuation from an ancient series of accelerations, which themselves, how-ever, have only just been discovered.

Limit line and approximate axis

Two parallel lines have been drawn obliquely across Fig. 11.3. The lower one is the 'equal-value line'. If at a million years before AD 2000, a process had a doubling time of a million years, its entry would lie exactly on this equal-value line; and so on. As a first approximation this line may be thought of as providing an axis for the whole series of entries. At least one can say that – with this particular form of plotting – most of the entries do lie quite near to this equal-value line.

The upper line is a limit or boundary line above which the recorded rates of acceleration (at various dates) do not in fact rise. The two lines assist in making the following statement:

At 10^t years before AD 2000 the recorded forefront doubling times are typically around 10^t years, and never less than $10^{(t-2)}$ years. Or in other words, at y years before AD 2000, the doubling times are typically around y years. And none of the doubling times are quite as short as a hundredth of y years.

Chapter 12

Aspects of number

The main function of this chapter will be to do a basic job of stock-taking. In the small part of the universe best known to us, the solar system, about how many are there of each of the nine main kinds of integrated natural entity? Inspecting the nine answers will suggest several general observations of wide validity.

Frequency estimates

Right from the beginning of an enquiry about the numbers of things, certain points stand out. One is that most of the mass of the solar system, nearly 99.9 per cent, resides in the Sun itself, the atomic composition of which has been reported in some detail. Another point is that the only living systems (or entities from intermediate entity upwards), which are known to occur in the solar system, are on Earth. The surface of the Earth is an extremely small habitat in comparison with the size of the whole solar system, but it is at least a place which we ourselves are able to observe closely. With regard to the photons emitted by the sun, it is proposed to include in the 'extended solar system' sufficient photons to bring the total number of fundamental particles up to the round number of 10^{60}. This will be discussed further in another chapter.

Thus we can have a firm value for the total number of fundamental particles in the extended solar system at Level 1. At the other end of the series at Level 9 we have a value of about 180 societies of sovereign states. This is much nearer to 10^2 than to 10^3, and for most purposes in this chapter will be treated as 10^2. But the estimates for the seven values between the two extremes are much more problematical. Even the number of complete hydrogen atoms in the sun is apparently less than the number of hydrogen atoms, as usually assessed. In an ordinary metal some of the outer electrons come apart from the atoms proper, and wander about between them. One speaks of

107

100+
Societies of sovereign States

Integrative level

100 000 000 000
Multifamily societies

10 000 000 000 000
One-mother family societies

10 000 000 000 000 000 000 000
Multicellular organisms

1 000 000 000 000 000 000 000 000 000
Cells with nuclei

100 000 000 000 000 000 000 000 000 000 000
Intermediate entities

10 000 000 000 000 000 000 000 000 000 000 000 000 000 000
Molecules

100 000 000 000 000 000 000 000 000 000 000 000 000 000 000 000 000
Atoms

1 000 000 000 000 000 000 000 000 000 000 000 000 000 000 000 000 000 000
Fundamental particles

Fig. 12.1. Tentative, pioneer estimates of the number of each of the nine kinds of integrated natural entity in the extended solar system. This list also serves to illustrate 'the law of the higher, the far far fewer'.

metallic bonding. Apparently some of the hydrogen of the sun exists in a condition similar to metallic bonding. And if a hydrogen atom's single electron is no longer constantly orbiting around one (or more than one) particular proton, that proton is no longer part of a Level 2 entity. For such an entity is an integrated group of two or more Level 1 entities.

Fig. 12.1 and Table 12.1 show the conclusions of this pioneer enterprise of taking stock of the constituent entities of the extended solar system. Though only tentatively, they provide an answer to the question of *how many*

Table 12.1. *The approximate* results of the basic job of 'stock-taking', together with the factors between the frequencies

Major integrative level	Kind of integrated natural entity	Estimated frequency in extended solar system	Factor between adjacent levels
9	Society of sovereign state	10^2	
			10^9
8	Multifamily society	10^{11}	
			10^2
7	One-mother family society	10^{13}	
			10^9
6	Multicellular organism	10^{22}	
			10^5
5	Cell with nucleus	10^{27}	
			10^5
4	Intermediate entity	10^{32}	
			10^{20}
3	Molecule	10^{52}	
			10^4
2	Atom	10^{56}	
			10^4
1	Fundamental particle	10^{60}	
Geometric mean of factors			$10^{7\cdot2}$
Or, omitting the 10^{20} geometric mean			$10^{5\cdot4}$

cells with nuclei, multicellular organisms, one-mother families, etc, *are* there on Earth, or in the solar system? There is no space here to recount the steps leading up to all the estimates cited.

The estimate of 10^{32} for the number of intermediate entities in the world endeavours to take account of the many small chromosomes of chloroplasts, mitochondria, plasmids and viruses. Gunnar Bratbak (1989) has studied the numbers of bacteriophages occurring in natural waters. In correspondence he suggests that there may perhaps be as many as 10^{27} to 10^{28} bacteriophages floating about loose in our natural salt and fresh waters, while the total of virus particles on the planet might be between 10^{28} and 10^{29}.

How many multicellular organisms are there, alive at any one time, on land and in the seas? On page 421 of *Sociobiology*, Wilson quotes an estimate that there are some 10^{18} individual insects. So perhaps the total of multicellular organisms is around 10^{22}.

The same series of frequencies, or numbers of things, is plotted in Fig. 12.2. Frequency histograms or frequency curves often show a more or less symmetrical or 'normal' distribution, with most of the individuals being in

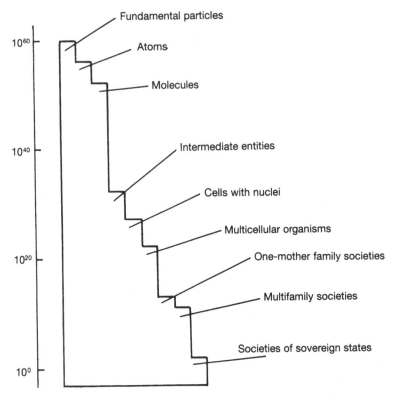

Fig. 12.2. Histogram showing the approximate frequencies of integrated natural entities in the extended solar system. This diagram uses a logarithmic axis of frequency.

the middle of the range. With such a symmetrical distribution the peak or
hump is in the middle, and thinning tails go off to left and right towards the
minimum or maximum values. But in Fig. 12.2 there is a conspicuously
asymmetrical or skew distribution. The peak is on the left, corresponding to
Level 1, so that in terms of statistics this can be described as a positively
skew frequency distribution. The reason for this kind of distribution is of
course built into the theory of major integrative levels, through the compositional and duality criteria. To give an example, there are obviously more
ordinary cells in multicellular organisms than there are multicellular organisms, while in addition there are unicellular organisms such as amoeba and
Chlamydomonas. Hence there are inevitably more Level 5 than Level 6 entities, and so on.

   The frequency estimates for all nine kinds of integrated natural entities lie
within the range from 10^2 to 10^{60}. The frequency of multicellular organisms

is taken as around 10^{22}, or ten thousand million million million. And it may be helpful to restate this estimate in terms of some small standard area. There are about 0.5×10^{15} square metres of surface area (land and sea) on our planet. So an estimate of 10^{22} multicellular organisms, in total, suggests an average of some 2×10^7 or 20 million multicellular organisms per square metre; and the true value quite likely does lie between 2 and 200 million. Perhaps an even clearer way of expressing the estimated average is as 2000 multicellular organisms per square centimetre of the surface of the Earth, land plus oceans. More needs to be known about the small creatures in various soils and in the sea. It is hoped that independent estimates from a number of investigators may be forthcoming.

Exponential trend

One of the striking features of Fig. 12.1 is the big gap between the frequencies of molecules and intermediate entities. We will return to that later. Another feature is that each step is of a hundredfold or more. But an important point which we should not ignore is that this histogram has been drawn using a logarithmic axis of frequency. The frequencies may also be plotted as in Fig. 12.3. This uses a straightforward arithmetic axis of frequency, where equal distances up the axis represent the addition of some constant number. Here most of the points lie virtually in the position of the zero baseline, which for that reason has not been drawn in. By joining the points one obtains an L-shaped plot. If the frequencies themselves had formed an arithmetic progression between the two extreme values – an arithmetic progression where from one term to the next a constant number is added (or taken away) – then they might have followed the dotted line model. Clearly the frequencies do not form an arithmetic progression, or anything like it. If one plots with a logarithmic axis as in Fig. 12.4 (which resembles Fig. 12.2), one sees that the general trend of the points is to form an exponential series, where each step from term to term is associated with a multiplication (or division) by a similar number. The points all lie within a linear band. An axis of relative scarcity (as discussed later) has also been marked in on the right of Fig. 12.4, and the trend of scarcity is of course similarly exponential.

Exceptional scarcity of living entities

The big step between molecules and intermediate entities in Fig. 12.2 requires further comment. Few readers will be surprised at the report that there are exceptionally few living entities in the solar system, in relation to molecules,

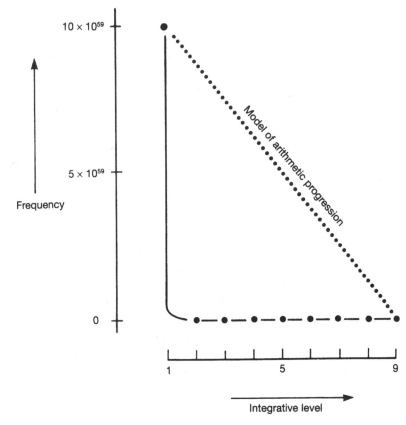

Fig. 12.3. Using an arithmetic axis of frequency, the frequency of each kind of entity is here plotted against integrative level. The actual series is clearly *not* an arithmetic progression.

etc. This may be formalised as: 'the law that life is very very rare'. As Table 12.1 reminds us, there may be some 10^{20} non-living molecules at Level 3, for each single living intermediate entity at Level 4.

It is well known that living things do happen to occur and flourish in the physical and chemical conditions provided by the surface of our planet, but they are presumably absent from the interior of the Earth, and apparently from everywhere else in the solar system. It is worth trying to quantify the smallness (in terms of mass) of the habitat of living things, on Earth, in relation to the whole solar system. The term biosphere can be used to refer to the layer of soil, water and air, around the planet, in which creatures live,

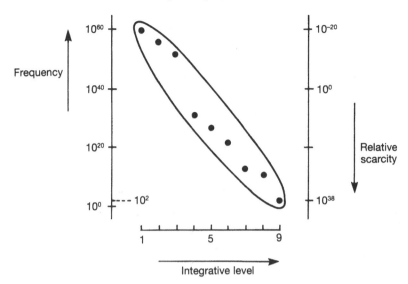

Fig. 12.4. Using now a logarithmic axis of frequency, it can be seen that the general trend of the nine frequency values is approximately exponential. The trend for relative scarcity is also exponential.

plus the living things themselves. But how much does this more or less spherical layer weigh? Or what fraction is its weight of the total weight of the Earth? Kaye and Laby's *Tables of Physical and Chemical Constants* (1966) is helpful. Consider the seas and oceans, the layer of soil on land where roots penetrate, plus man and his artifacts and all other living organisms, together with the air in which planes and birds fly. In that list of the parts of the habitat or environment of organisms, including the organisms themselves – roughly the biospheres – the weight of the seas and oceans predominates. One needs to multiply the mass of this habitat by about $10^{3.5}$, or by a factor of 3000+, to obtain the total mass of the Earth. Then multiplying the mass of the Earth by about $10^{5.5}$ brings one up to the mass of the solar system. And $10^{3.5}$ times $10^{5.5}$ equals 10^9, that is a thousand million or one billion.

Hence the mass of the habitat of living things, on Earth, is only about a thousand-millionth of the mass of the whole solar system. That certainly helps to explain the scarcity of living things. But there are apparently some 10^{20} more molecules than intermediate entities. Even if that is divided by 10^9, the substantial factor of 10^{11} still remains. However inspection of Table 12.2 shows that there are factors as great as about 10^9 between multicellular

Table 12.2. *The relative scarcities, as estimated, of the different kinds of integrated natural entities*

Major Level	Relative scarcity of entities	Group
9	10^{38}	
8	10^{29}	
7	10^{27}	Social and biological entities
6	10^{18}	
5	10^{13}	
4	10^{8}	
3	10^{-12}	
2	10^{-16}	Non-living entities
1	10^{-20}	

organisms and unit families, as well as between multifamily societies and nations.

Steeply progressive scarcity

One interesting way of expressing the relative scarcity of each category in the extended solar system is to divide its frequency into 10^{40}. This gives the values of an index of relative scarcity, as listed in Table 12.2. All living entities have values greater than unity, in contrast to the 'below the line' entities of the physical range, which are so much commoner.

The values of relative scarcity, for entities of each of the nine major integrative levels, may be plotted as Fig. 12.5. This particular presentation draws attention to the way the values seem to hold apart, in two separate groups. (One reason is of course that only a thousand-millionth of the solar system provides a suitable habitat.) But each of the two groups has a steep and similar exponential trend upwards. Within the two groups the geometric mean factor from term to term is about $\times 10^{5.4}$, as one proceeds up the natural hierarchy (Table 12.1).

Clearly then, among the inanimate as well as among the animate entities, we observe steeply progressive scarcity. There are often 100 000 or so entities at one level, for every single entity at the level above. Recollecting the old saying 'the higher the fewer', we might perhaps now speak of 'the law of higher the far far fewer'.

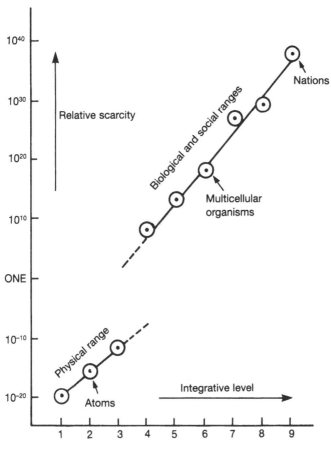

Fig. 12.5. Relative scarcity is plotted against integrative level, showing an obvious discontinuity between the inanimate and animate groups. Within each group there is a steep exponential trend, but on present data the two slopes are slightly different.

Chapter 13
Aspects of mass

Towards the end of Chapter 6 we had a logarithmic diagram (Fig. 6.2) which showed the *range of mass* of each of the first six kinds of integrated natural entity, i.e. those in the physical range plus those in the biological range. For each level there was an approximate division into main and subsidiary portions of the total range of mass. Now in Fig. 13.1 the same information is repeated but entries have also been made for societies at Levels 7, 8 and 9. Such societies occur only in truly social species, where the adult actively aids the young after hatching or birth. At Level 7 there are the one-mother families, ranging from those of small insects to great whales; and the same range includes all one-kitchen human households, both private and institutional. At Level 8 there are the multifamily domiciliary societies. These also occur among insects and whales, for instance. The mass of the population of some human settlements is very great indeed. Then at Level 9 we have the societies of sovereign states, occurring in *Homo sapiens* only.

The entry in Fig. 13.1 for the societies of sovereign states uses data from a few years ago, when Iceland was still the nation with the smallest population. China, with the largest population, had at least three thousand times as many people as Iceland. The size of the factor between them does of course determine the length of the entry for societies of sovereign states in Fig. 13.1. This entry happens to be a comparatively short one. Inspection shows that atoms are the only kinds of entity to have a smaller range between lightest and heaviest, while for most of the categories there is a far greater range. Both atoms and nations have a range of mass of more than a hundredfold. But the geometric mean of the nine ranges, taken all together, is over a million-million-fold.

Besides noting the long ranges for most kinds of entities, there is the corresponding phenomenon of the overlapping of the ranges. Many multifamily societies can be matched, in weight, by various single multicellular

116

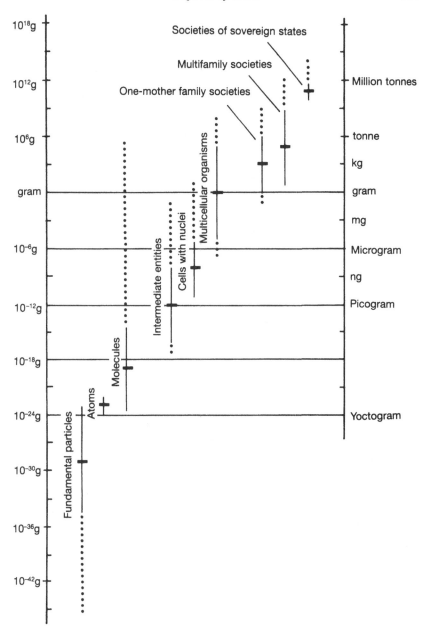

Fig. 13.1. The main and subsidiary ranges of mass of the nine kinds of integrated natural entity. This is an extension of Fig. 6.2. The 'characteristic mass' is also marked, by a short line bisecting the main range. Several million-fold bands have been marked in.

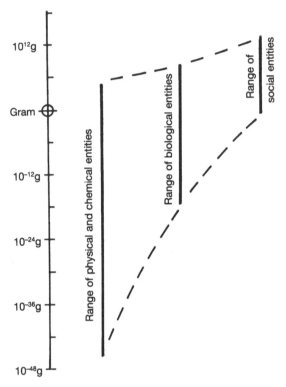

Fig. 13.2. An illustration of the decrease in range of mass as one ascends, from group to group, up the natural hierarchy. On the left is the estimated range from the lightest photon in motion, to the heaviest molecule. In the middle is the range from the lightest intermediate entity to the heaviest multicellular organism. On the right, the range from the lightest unit family to the heaviest society of sovereign state.

organisms. And an integrated natural entity of one microgram, as one can see from Fig. 13.1 at a glance, could be either a member of Level 3, 4, 5 or 6.

Why bother about the property of weight or mass? The complexity, influence and power of a human society, or a single person, is difficult to express quantitatively. With electrons and photons one cannot even be precise about such properties as length, width or volume but it so happens that the basic property of mass can be quantified for the whole range of integrated natural entities. So for wide-ranging comparisons, mass is one of the very few properties which can be used. It is certainly a novelty to witness any such gravimetric review, as in Fig. 13.1. The total ranges of mass, for each kind of entity, give a reasonably accurate impression, though possibly a few of the scarce outlying members of some categories may have been inadvertently omitted.

Comparing the physical, biological and social groups of ranges

Despite atoms having such a small total range, it can be shown, from the data as plotted, that the physical group of entities have a geometric mean (of their three total ranges of mass) of $10^{17.7}$, while the biological group have a smaller mean of $10^{14.7}$, and the mean for the social range is only $10^{7.7}$, or we could round off these three values as 10^{18}, 10^{15} and 10^{8}. Hence we may record a general decrease in the range of mass as one proceeds from group to group up the natural hierarchy.

Another way of studying the trend of ranges of mass is to lump all three kinds of entity in the physical range together; and so on. Thus from the lightest fundamental particle in motion, as plotted, to the heaviest molecule, the factor is about 10^{50}. Then in the biological group, the factor from the lightest intermediate entity to the heaviest multicellular organism has a much lower value of about 10^{26}. In the social group, from the lightest unit family to the heaviest nation the factor is only about 10^{15}. This is a further illustration, as in Fig. 13.2, of the general decrease in range of mass, as one proceeds up the natural hierarchy.

Characteristic mass

For each of the nine ranges in Fig. 13.1 a main range, containing most of the better-known members, has been marked by a continuous line. For the rest of the total range, containing fewer members, dotted lines have been used. But this rough and ready distinction, though useful, has no precise statistical basis. Having selected each main range in a somewhat arbitrary manner, and marked it by a continuous line, the middle point of this line was then marked in, as the 'characteristic mass' of the type of entity concerned. If more complete information were available, one might prefer to use the commonest or modal value along this logarithmic scale; but at present the characteristic value seems the best available substitute.

The nine values for characteristic mass are listed in Table 13.1. It may be mentioned that the round-number nature of the values is due to the main ranges having originally been arbitrarily drawn so that in each case the middle point was at an exact multiple or submultiple of ten. The values of characteristic mass range from 10^{11} grams down to 10^{-29} gram. Hence the extreme values differ by a factor of 10^{40}.

Table 13.1 also lists the eight factors between the successive values of characteristic mass. This series of factors is rather similar to that in Table 12.2, with the difference that here there is no outsize factor, like that of 10^{20},

Table 13.1. *Estimates of the value of characteristic mass, at each level,*
and of the factors separating them

Major integrative level	Kind of integrated natural entity	Estimate of characteristic mass (grams)	Factor between adjacent levels
9	Society of sovereign state	10^{11}	
			10^6
8	Multifamily society	10^5	
			10^2
7	One-mother family society	10^3	
			10^3
6	Multicellular organism	10^0	
			10^8
5	Cell with nucleus	10^{-8}	
			10^4
4	Intermediate entity	10^{-12}	
			10^7
3	Molecule	10^{-19}	
			10^4
2	Atom	10^{-23}	
			10^6
1	Fundamental particle	10^{-29}	
Geometric mean of factors			$10^{5.0}$

between molecules and intermediate entities. Here there is a smooth and steady progression between the physical and biological ranges. The geometric mean of the factors here is $10^{5.0}$, compared with $10^{5.4}$ of Table 12.1. Indeed the round-number geometric mean of $10^{5.0}$ can here be obtained rather easily as the eighth root of 10^{40}. This factor of 10^5 implies that the value of the characteristic mass, at one level, tends to be around a hundred thousand times the characteristic mass of entities at the level next below.

It is rather interesting to plot the values of characteristic mass against the numbers of the levels concerned. Fig. 13.3 uses an *arithmetic* scale of mass. The result is an extreme example of a 'reversed-L curve'. However, using a logarithmic axis of mass (Fig. 13.4) shows that the values tend to behave as an exponential series. None of the entities depart very far from the line which has been drawn across. So making use of this line, we can state concisely that the characteristic mass, in grams, of entities of the nth integrate level, is approximately $10^{(5n-34)}$.

This chapter has embarked upon a novel gravimetric review of the masses of all the integrated natural entities which are members of the nine major

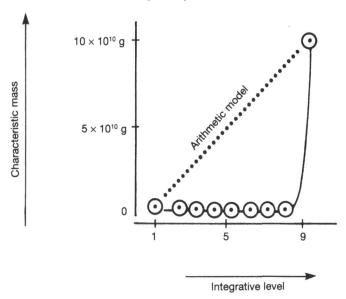

Fig. 13.3. Integrative level plotted against characteristic mass, using an arithmetic axis.

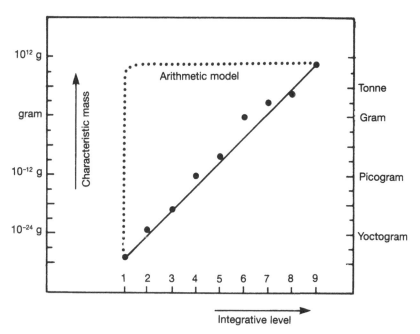

Fig. 13.4. The same data as Fig. 13.3 but using a logarithmic axis of mass. This shows an obviously exponential relationship between characteristic mass and integrative level.

integrative levels. Perhaps the following are the main conclusions which stand out.

At each level the range of mass is considerable. Even for atoms and for nations it is more than hundred-fold. But it is usually much greater than that. Indeed the geometric mean of all the nine ranges is more than million-million-fold.

Correspondingly there is an enormous amount of overlapping (Fig. 13.1). For instance an integrated natural entity of one microgram might be a member of Level 3, 4, 5 or 6.

The characteristic mass of one major integrative level is often around 10^5 times that of the level next below. Characteristic mass rises exponentially as one proceeds up the natural hierarchy (Fig. 13.4).

Chapter 14

Positive skewness

At a given integrative level are there only a few lightweight members, many middleweight members, and only a few heavyweights? That is the kind of symmetrical or 'normal' distribution which is often encountered or assumed in the study of statistics. Or will some or all of the integrative levels show some other kind of distribution?

Social range

Level 9: Societies of sovereign states

For plotting the diagram of Fig. 14.1 the information used was from the 1971 data sheet of the Population Reference Bureau in Washington. The conspicuous result is that about three-quarters of all states consist of less than 20 million people. These include for instance Sweden, Switzerland, Sudan, Somalia, Senegal, Syria, Saudi Arabia, Singapore and El Salvador, to mention just a few.

Most of the other quarter have populations between 20 and 200 million, including for instance the United Kingdom, France, Germany, Spain, Italy, Egypt, Canada and Japan. Only the United States, Russia, India and China had populations in the range beyond 200 million. So the long thin tail to the right of Fig. 14.1 is hardly to be distinguished from the zero baseline; and for that reason the baseline itself has not been drawn in. (In recent years the population of China has of course grown to more than 800 million, indeed by 1990 to more than a thousand million; and India itself might reach that population by the year 2000.)

The vertical axis shows the relative frequency or number of nations along different sections of the horizontal scale. There is an enormous hump at the left-hand end of the range, close to the minimum value. So here the

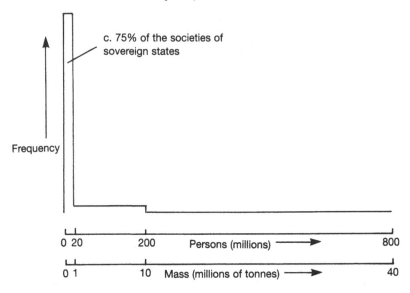

Fig. 14.1. The smaller and larger sovereign states in the world are grouped (using 1971 data) according to the number of people they contain, and similarly according to the total weight of their populations. The horizontal axes are of the number of persons per sovereign state and their mass. The axis of relative frequency, used here for this frequency histogram, is a straightforward arithmetic axis as are the frequency axes in Fig. 14.2 and 14.3.

commonest number of people per state is conspicuously *not* near the middle of the range, at around 400 million. The frequency distribution is a one-sided or skew distribution. And with the commonest value occurring near to the minimum end, it can be technically described as a positively skew distribution.

So far we have been dealing with numbers of people. But if we use a convention of taking the average mass per person as 50 kilograms, as on the lower horizontal axis, then the same histogram can be used to show the distribution of nations with regard to their mass. (The true average is in fact less than 50 kg per person.) The fact that the average weight per person is a little different in different countries will not affect the general picture obtained. Clearly the frequency distribution of nations with regard to mass is positively skew. About three quarters of the nations have a mass of less than a million tonnes (or 10^{12} grams), while the mass of the population of China is many times greater.

Level 8: Multifamily societies

Moving from Level 9 to Level 8, it is interesting to plot the frequency with regard to mass, of the societies of local authority districts in the United King-

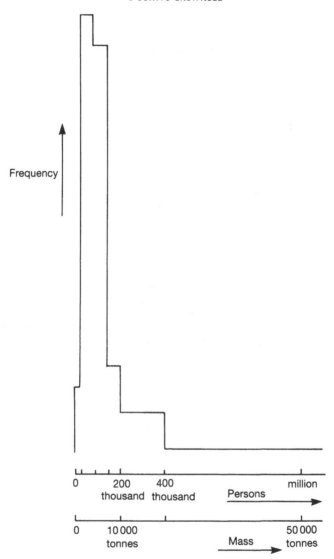

Fig. 14.2. As Fig. 14.1 but concerning local authority districts of the United Kingdom. The people in each such district constitute one administrative unit and here again the units are grouped according to the size and weight of their population.

dom. Fig. 14.2 uses the data from *Whitaker's Almanack* for 1976, and the same two horizontal axes are used as for Fig. 14.1. Here again the hump is near the minimum value. The distribution is again strongly positively skew. In a sample of 300 of our geographically defined settlements (Pettersson 1960b) the frequency distribution was positively skew even if they were

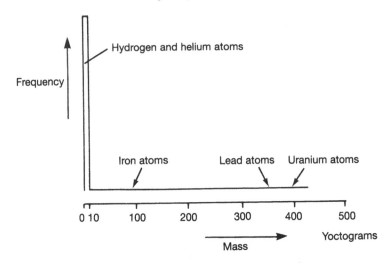

Fig. 14.3. On an arithmetic axis the relative frequency, with regard to mass, of atoms (in the sun and the solar system). Helium as well as hydrogen atoms weigh less than 10 yoctograms. Compared with these the numbers of all other kinds of heavier atoms are negligible, so far as the scale of this particular diagram is concerned. But the position of atoms of iron, lead and uranium (at 395 yoctograms) is shown; and the plot is extended to include several man-made elements.

grouped according to logarith of population. With a straight forward arithmetic axis the distribution would be even more skew.

So far we have only considered the multifamily domiciliary societies in one country, and societies only of our own species. In fact the most numerous multifamily societies are those of various insects, especially ants and termites. These are not only very numerous, but also of relatively slight mass. So with humans, ants, termites and other species all taken together, it is clear that the frequency distribution of multifamily societies, with regard to mass, is very positively skew indeed.

Physical range

Atoms

Another level for which relatively precise information is available is Level 2, that of atoms. The handbook of *Tables of Physical and Chemical Constants* (originated by Kaye and Laby 1966) contains a well-known table of the estimated relative abundance of atoms of different elements, both for the Sun and for the universe in general. For the present purpose the data for the Sun will do equally well for the whole solar system (Fig. 14.3). The great hump on the left

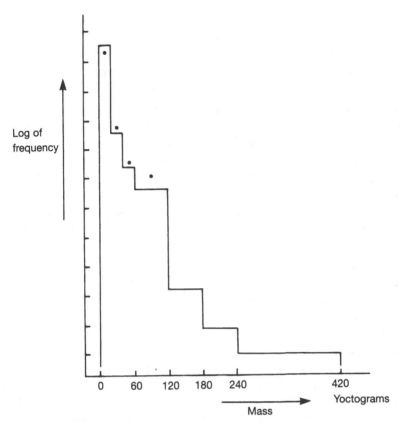

Fig. 14.4. Using a logarithmic axis of frequency, the main histogram shows the estimated relative frequency of atoms in the universe. (Each vertical interval represents a 10-fold change.) In four categories, the dots show the relative frequencies of atoms in the sun.

contains all the hydrogen and helium atoms, in comparison with which there are very few indeed of any heavier atoms. So having switched from enormous social entities, down to the level of ultramicroscopic atoms, we *again* find that the frequency distribution at this major level, with regard to mass, is strongly positively skew. (The fact that not all hydrogen 'atoms' in the Sun are actually complete atoms does not affect the simple histogram of Fig. 14.3.)

Since so little of the total information available shows in Fig. 14.3, in Fig. 14.4 the unusual step is taken (as in Fig. 12.2) of using a logarithmic axis to express relative frequencies. The main histogram is for number of atoms, in different bands of mass, in the universe in general. For the Sun, though not for the whole solar system, estimates are available for the first four bands (not others) and these frequencies have been indicated by four dots on the same diagram.

Fundamental particles

Turning to the fundamental particles of Level 1, there are in the solar system slightly more protons plus neutrons at 1.67 yoctograms than electrons at about 0.001 yoctograms. The hyperons, which have been known for some years, go up to 2.4 yoctograms in mass. But in 1976 came news of a pair of much heavier fundamental particles. In fact Richter and Ting received a Nobel Prize in 1976 for their discovery of the two psi particles, the heavier of which has a mass of about 4 atomic mass units, or daltons, so that its metric mass works out at 6.6 yoctograms. And in 1989 the Z° particle was discovered. As mentioned in Chapter 4 this is much heavier still.

Even before the discovery of the Z° particle, the frequency distribution of the fundamental particles of the solar system could be clearly seen to be positively skew also. Fig. 14.5 shows several versions of the histogram, according to what limits one uses to define the solar system. If one considers the solar system to go no further out than the orbit of the farthest planet, Pluto, the frequency distribution is as in (a) with protons and neutrons slightly more numerous than electrons, etc. But should one include some of the photons (in motion), given out by the sun, some time ago, which have by now moved beyond Pluto's orbit? Histogram (b) includes as many photons as electrons. Histogram (c), however, is for an 'extended solar system' (as in Chapter 12) which includes sufficient photons from the sun to bring the total number of fundamental particles up to the round number of 10^{60}. All these diagrams show the fundamental particles of the solar system as having a positively skew distribution with regard to mass. (There are now reports of a planet still further from the Sun than Pluto being discovered.)

Diagrams (a), (b) and (c) treat all the particles of between 0 and 1 yoctogram as in one group; and all the particles of between 1 and 2 yoctograms as in one group. However, the majority of photons in motion, as well as electrons, have a mass which is only a very small fraction of one yoctogram. So replotting the data from (c), Fig. 14.5 (d) can give us a more realistic picture. It shows the same sort of exaggeratedly lopsided distribution which is encountered among atoms and among the societies of sovereign states (Figs. 14.3, 14.1), although the hump on the left in (d) could be drawn even thinner still.

It may be wondered whether Fig. 14.5 (c, d) may also be a description of the frequency distribution of the fundamental particles of the universe in general. This comment was already published in 1976. Barrow and Silk (1980) mention an estimate that there are some 10^8 photons in the universe for every single nucleon. (The term nucleon includes both protons and neutrons.) So whether one (a) chooses to regard photons as being without mass, or (b) chooses to con-

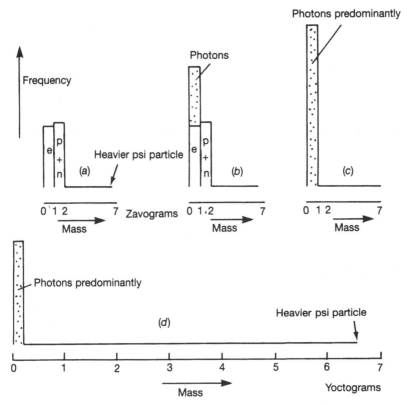

Fig. 14.5. The frequency distribution, with regard to mass, of fundamental particles – though only (*a*) and (*b*) use the same scale of frequency. Some of the histograms refer to the extended solar system (see text). Each photon, in motion, is normally thought of as possessing a small but definite mass. The above plots would, however, be equally true if one thought of the 'rest mass' of the photon as being zero. These diagrams predate the discovery of the Z° particle.

sider the small but finite mass of photons in motion, the universe is indeed now reported as having – emphatically – a positively skew frequency distribution of fundamental particles with regard to mass.

General conclusions

Evidence has been presented about the one-sided distribution, with regard to mass, of the entities at each of four major integrative levels. In each case the commonest values are much nearer to the minimum than to the maximum value. For the other five major integrative levels, there are no comprehensive tables but supplementary information may be gained from Fig. 13.1.

Some explanation may be desirable at this point. Looking at Fig. 13.1, we can see the distance up the right-hand axis from, for instance, a gram to a kilogram. This represents a factor of × 1000, or going from kilogram to gram represents a factor of a thousandth. From the level of gram to the level of tonne represents a factor of a million, or × 10^6. There are several million-fold bands, marked on the figure between the horizontal lines. A million is approximately 2^{20}, thus going up from one of the horizontal lines one-twentieth of the distance to the next horizontal line above, represents going up by a factor of × 2, i.e. going up to twice the weight one started from; or going up by × $10^{0\cdot3}$. Going up by that small distance on any other part of the figure will similarly mean a movement to twice the weight one starts from.

In Fig. 13.1, the main range of mass for each of the nine kinds of entity is represented by a continuous line. Above it there is a dotted line indicating the subsidiary range which includes the relatively scarce heavier entities of each kind. This upper dotted line for ordinary cells (as drawn) represents a factor of about a million-fold. Some of the other upper dotted lines are nearly as long or longer. Even for atoms and fundamental particles the rather short upper dotted lines are longer than that very short distance which would represent a two-fold movement to a weight twice that of the heaviest end of the main range. All the other seven kinds of entity have upper dotted lines which are longer than those for fundamental particles and atoms. Hence in all the nine cases the heaviest entity is more than twice as heavy as those at the heavier end of the main range of mass (and usually many times heavier). It may further be noted that in each case the great majority of the entities along their main range of mass (unbroken line) are much lighter than those at the heaviest end of the main range. Therefore Fig. 13.1 itself serves to indicate that the frequency distribution with regard to mass is strongly positively skew at each of the nine major integrative levels.

With the teaching of statistics giving a prominent place to samples which do have a symmetrical or normal distribution, this pervasive and emphatic positive skewness stands out in clear contrast.

Chapter 15
Quantitative conclusions

It now seems appropriate to review the main quantitative conclusions of the foregoing chapters. Most of the conclusions involve entities of all three ranges, the physical, the biological and the social. The conclusions will be listed according to chapter of origin, and the relevant figures and tables will be mentioned to facilitate reference back.

Major integrative levels (Chapter 2)

By making use of the compositional criterion and the duality criterion, it is found that the number of major integrative levels – into which the integrated natural entities around us can be grouped – is no more nor less than nine. There are three levels in the physical range, three in the biological range and three in the social range (Table 2.2, Fig. 1.3, and for the duality criterion Fig. 2.1 and Fig. 2.2).

Acceleration in evolution (Chapter 10)

The period of time before entities of a higher integrative level have emerged from the biological or social level below has, in general, decreased with the advance of time (Fig. 10.1).

When the estimated mass of innovatory entities is considered, they are found to form an exponential series, or at times to increase faster than in a simple exponential manner. The doubling time has decreased from about 75 million years, down to about 12 million years (Fig. 10.4).

Further accelerations (Chapter 11)

The increasing numbers of people in the world is found, *numerically*, to approximately continue the same acceleratory trend as evinced by the earlier evolutionary acceleration (Fig. 11.1).

Since 10 000 years ago the population of the world's largest settlement (village, town and then city) has increased in an acceleratory and approximately exponential manner, with a doubling time (until recently) of about 600 years (Fig. 11.1).

The published results of several scholars show – when put together – that over the past 30 000 years the rates of acceleration, in a variety of forefront innovatory cultural developments, among humans, have fairly consistently increased. Cultural changes have been getting faster. The doubling times have shortened for instance, from 100 000 years down to one year (Fig. 11.2).

This acceleration in cultural innovation has *also* more or less continued the same trend as the demographic and earlier evolutionary accelerations (Fig. 11.3). The overall range of doubling times is from about 75 million years down to one year.

The broad similarity between cultural increase and demographic increase is very unlikely to continue indefinitely. Indeed the increase in world population may soon cease altogether.

Aspects of number (Chapter 12)

Estimates have been made of the number of integrated natural entities, of each major level, in the 'extended' solar system. The numbers range from 10^{60} for Level 1, to rather more than 10^2 for Level 9 (Fig. 12.1 and Table 12.1).

Inspection of Fig. 12.2 and Table 12.1 shows the tremendous discontinuity with regard to frequencies, between the physical and the biological ranges. There are some 10^{20} inanimate molecules, to every single living intermediate entity. One might perhaps speak of 'the law that life is very very rare'. But the scarcity of living things is of course partly due to such a small portion of the solar system providing satisfactory physical and chemical conditions.

If one omits that 10^{20}, the geometric mean of the other factors cited in Table 12.1 is $10^{5.4}$. Thus there are often several or many thousands of entities at one level, for every single entity at the level above.

There is seen to be a distinctly exponential trend, if frequency of entity (or relative scarcity) is plotted against integrative level (Figs. 12.4, 12.5).

Aspects of mass (Chapter 13)

It is interesting to compare the range of mass at a certain level (the range between lightest and heaviest entity) with ranges at other levels. Even for atoms at Level 2, and for the societies of sovereign states at Level 9, the range is more than one hundred-fold (Fig. 13.1).

However, the geometric mean of the nine ranges, taken all together, is more than a million-million-fold.

As Fig. 13.1 shows, there is a very considerable overlapping of the ranges. Thus an integrated natural entity weighing one microgram might be a member of Level 3, 4, 5 or 6.

There is a striking decrease in range of mass as one passes from the physical group to the biological group and then to the social group (e.g. Fig. 13.2).

The characteristic mass of each of the nine levels has been estimated. The characteristic mass of one level tends to be around 10^5 times that of the level immediately below (Table 13.1).

Characteristic mass shows a clear exponential trend when plotted against integrative level (Fig. 13.4).

Positive skewness (Chapter 14)

With regard to their occurrence in the solar system, it has long been widely known that the majority of atoms are situated nearer to the minimum end rather than to the maximum end of their range of mass. That is atoms have a positively skew distribution with regard to mass (Fig. 14.1). But it has now been discovered that members of *all* the nine major levels have a positively skew distribution.

For the universe as a whole, it is now reported that fundamental particles as well as atoms have a positively skew distribution with regard to mass. Thus positive skewness is a very pervasive feature indeed.

Conclusion

We have just listed some twenty quantitative conclusions, of wide validity. Nearly all of them involve entities in all three ranges, the physical and biological as well as the social range.

The main conclusions might, if one wished, be stated as so many laws. They would, however, be 'descriptive' laws, and not 'predictive' laws such as those of Boyle and Ohm.

This book, which deals with *all* the nine levels of the natural hierarchy, has given the opportunity for these widely based comparisons and conclusions to emerge. At the same time it has been helpful in binding up the different individual sciences into a corporate and coherent whole.

Bibliography

Admiralty Handbook of Wireless Telegraphy, London: H.M.S.O.

Barrow, J. D. & Silk, J. (1980). The structure of the early universe. *Scientific American*, **242**, 4, 98–108.

Bender, B. (1975). *Farming in Prehistory*, London: John Baker.

Blythe, R. (1969). *Akenfield*, London: Allen Lane.

Bratbak, G. *et al.* (1989). High abundance of viruses found in aquatic environments. *Nature*, **240**, 467–8.

Carr-Saunders, A. M., Jones, D. C. & Moser, C. A. (1958). *A Survey of Social Conditions in England and Wales*, London: Oxford University Press.

Childe, V. G. (1942). *What Happened in History*, London: Penguin Books.

Cohen, D. (1965). *The Biological Role of the Nucleic Acids*, London: Edward Arnold.

Cunliffe, B. (1974). *Iron Age Communities in Britain*, London: Routledge & Kegan Paul.

Daniel, G. (1968). *The First Civilizations*, London: Thames & Hudson.

Danloux-Dumesnils, M., translated by Garrett, A. and Rowlinson, J. S. (1969), *The Metric System*, London: The Athlone Press.

Douglas, J. (1975), *Bacteriophages*, London: Chapman and Hall.

Durrell, G. (1964). *A Zoo in my Luggage*, London: Penguin Books.

Ehrlich, P. R. & Ehrlich, A. H. (1970), *Population, Resources, Environment: Issues in Human Ecology*, San Francisco: W. H. Freeman.

Ehrlich, P. R., Ehrlich, A. H. & Holdren, J. P. (1978), *Ecoscience: Population, Resources, Environment*, San Francisco: W. H. Freeman.

Foskett, D. J. (1978), The theory of integrative levels and its relevance to the design of information systems, *Aslib Proceedings*, **30**(6), 202–8.

Fossey, D. (1983), *Gorillas in the Mist*, London: Hodder and Stoughton.

Gamkrelidze, T. V. & Ivanov, V. V. (1990), The early history of Indo-European languages, *Scientific American*, **262**, 3, 82–9.

Greenberg, J. H. & Ruhlen, M. (1992), Linguistic origins of native Americans, *Scientific American*, **267**, 5, 60–5.

Hinde, R. (1974), *The Biological Basic of Human Social Behaviour*, Maidenhead: McGraw-Hill.

Jolley, J. L. (1973), *The Fabric of Knowledge*, London: Duckworth.

Kaye, G. W. C. & Laby, T. H. (1966), *Tables of Physical and Chemical Constants*, London: Longman.

King, J. A. (1955), *Social Behaviour, Social Organization, and Population Dynamics in a Black-tailed Prairiedog Town in the Black Hills of South*

Dakota, (123 pp), Contributions from the Laboratory of Vertebrate Biology no. 67. Ann Arbor: University of Michigan Press.

Lack, D. (1965), *The Life of the Robin*, London: Witherby.

Laslett, P. (1968), *The World we have Lost*, London: Methuen.

Leakey, R. E. & Lewin, R. (1978), *Origins*, London: Macdonald and Jane's.

Mackie, E. W. (1977), *Science and Society in Prehistoric Britain*, London: Paul Elek.

Ministry of Defence (1987), *The British Army*, London: Her Majesty's Stationery Office.

Monod, J. (translated from the French by A. Wainhouse) (1972), *Chance and Necessity*, London: Collins.

Morgan, T. H. (1911), Random segregation versus coupling in mendelian inheritance, *Science*, n.s. **34**, 384.

Morley, D. W. (1953), *The Ant World*, London: Penguin Books.

National Physical Laboratory (1991), *Units of Measurement*, a broadsheet, Teddington: National Physical Laboratory.

Needham, J. (1931), *Chemical Embryology*, Cambridge: Cambridge University Press.

Needham, J. (1943), *Time, the Refreshing River*, London: Allen and Unwin.

Needham, J. *et al.* (1954–), *Science and Civilisation in China*, Cambridge: Cambridge University Press.

Noyce, R. N. (1977), Microelectronics, *Scientific American*, **273**, 3, 62–9.

Palmer, L. S. (1957), *Man's Journey through Time*, London: Hutchinson.

Parker, R. (1975), *The Common Stream*, London: Collins.

Parsons, J. (1971), *Population versus Liberty*, London: Pemberton Books.

Pauling, L. (1960), *The Nature of the Chemical Bond*, Third edition. New York: Cornell University Press.

Pettersson, M. L. R. (1956), Diffusion of a new habit among greenfinches, *Nature*, **177**, 709–10.

Pettersson, M. L. R. (1959a), Greenfinch feast mass-observed – with map. *The Observer*, 2 August.

Pettersson, M. L. R. (1959b), The despoliation of *Daphne* and the relative success of pest control measures – a report of a co-operative investigation, *Journal of the Royal Horticultural Society*, **84**, 379–81.

Pettersson, M. L. R. (1960a), Intraspecific differentiation: biological and social aspects, *Nature*, **186**, 431–2.

Pettersson, M. L. R. (1960b), Increase of settlement size and population since the inception of agriculture, *Nature*, **186**, 870–2.

Pettersson, M. L. R. (1960c), Cultural diffusion in other animals and in man, *Biology and Human Affairs*, **25**, 24–9.

Pettersson, M. L. R. (1961), The nature and spread of *Daphne*-eating in the greenfinch, and the spread of some other new habits, *Animal Behaviour*, **9**, 114.

Pettersson, M. L. R. (1961), Society means ... what?, *New Scientists*, **9**, 355.

Pettersson, M. L. R. (1962), The stages of social evolution of man and his ancestors, *Man*, **62**, 103–4.

Pettersson, M. L. R. (1964), On the convenience of more thorough metric usage in the biological and allied sciences, *Journal of Theoretical Biology*, **6**, 217–43.

Pettersson, M. L. R. (1976), *Mainstream Hierarchy*, London: Brunel University.

Pettersson, M. L. R. (1978a), Acceleration in evolution, before human times, *Journal of Social and Biological Structures*, **1**, 201–6.

Pettersson, M. L. R. (1978b), Major integrative levels and the *fo-so* series, *Aslib Proceedings*, **30**(6), 215–37.

Pettersson, M. L. R. (1979), Vertical taxonomy: for certain social, biological and physical structures, *Journal of Social and Biological Structures*, **2**, 255–67.

Pettersson, M. L. R. & Pritchard, J. S. (1988), *Daphne* and the greenfinches, *Journal of the Royal Horticultural Society*, **113**, 245.

Price, D. de Solla (1956), The exponential curve of science, *Discovery*, **17**, 240–3.

Renfrew, C. (1971), Carbon 14 and the prehistory of Europe, *Scientific American*, **225**, 4, 63–72.

Renfrew, C. (1989), The origins of Indo-Europeans languages, *Scientific American*, **261**, 4, 82–90.

Richmond, M. H. & Smith, D. C. (eds.) (1979), *The Cell as a Habitat*, London: The Royal Society.

Roberts, J. M. (1976), *The History of the World*, London: Hutchinson.

Scarre, C. (ed.) (1988), *Past Worlds, The Times Atlas of Archaeology*, London: Times Books.

Schaller, G. B. (1965), *The Year of the Gorilla*, London: Collins.

Thomson, D. (1978), *England in the Nineteenth Century*, London: Penguin Books.

Tinbergen, N. (1965), *Social Behaviour in Animals*, London: Chapman and Hall.

Toffler, A. (1970), *Future Shock*, London: Bodley Head.

Turner, C. G. (1989), Teeth and prehistory in Asia, *Scientific American*, **260**, 2, 70–77.

Watson, J. D. & Crick, F. H. C. (1953), Molecular structure of nucleic acids: a structure for deoxyribose nucleic acid, *Nature*, **171**, 737–8.

Watson, J. D. (1970), *The Double Helix*, London: Penguin Books.

Werskey, G. (1978), *The Visible College*, London: Allen Lane.

Whitaker's Almanack 1976, London: Whitaker.

Wilson, E. O. (1975), *Sociobiology*, Cambridge, Mass.: Harvard University Press.

Index

acids, 33
adenine, 38
adenosine triphosphate, (ATP) 55
administrative areas, settlements and,
    75–9
Africa: mountain gorillas, 65–6, 70
aggregational natural entities, 1
agriculture, 69, 74, 75
    and societies, 86
algae, 10, 54, 56
aluminium atoms, 18
amino acids, 51
amoeba, 10, 11, 53, 57
amphibians, 90
Anatolia, 82–3
antiprotons, 27
ants, 67, 126
argon, 31
army hierarchies, 4, 7
Arrhenius, S., 34
artifacts
    complexity of, 102
    household, 71, 72, 73
artificial entities, integrated, 1
Asia
    agriculture, 75
    societies in, 83
asymmetrical distribution, 110, 123–30
atomic mass units, 23, 28, 29
atoms, 1–3, 4–6, 9, 26–7
    duality of behaviour, 31–2
    number in solar system, 107–10
    size and mass, 17, 18, 20, 28–30, 58
        characteristic mass, 120
        gravimetric triangles, 36–7, 59, 60,
            88
        skew mass distribution, 126–7
    see also fundamental particles
Avogadro's constant, 28–30

bacteria, 14, 42, 53–4, 55, 73

as intermediate entities, 2
    sizes of, 17–18
    volume of, 23
bacteriophages, 54, 55, 109
Barrow, J.D., 128
baryons, 27
bases, 33
bees, 62, 64–5, 68
Bender, B., 82
Bible, 69
biomass, see mass
biosphere, mass 112–13
bird societies, 61, 62, 63, 68
black-tailed prairie dog, 66–7
blood plasma, 12, 14
blue-green-algae, 54
Blythe, R., 78
Bragg, W.L., 34
brain, 43
    acceleration of size, 97
Bratbak, G. 109
Britain
    agriculture in, 75
    geographical/administrative entities, 75–9,
        124–5
    households in, 73
    mass, 87–8
    neolithic societies, 83
    tribal societies, 83–4
British Army: hierarchy, 7
brown algae, 56

capital cities, 81–2
carbohydrate synthesis, 55
carbon atoms, 18
    isotopes of, 30–1
    mass of, 23, 28, 29
        gravimetric triangles, 36–8, 59, 60
carbon dioxide, 42
Carr-Saunders, A.M., 72–3

138